13th Ranshofen Light Metal Days 2024

IOP Conference Series: Materials Science and Engineering
Volume 1315

Saalfelden, Austria
26-27 September 2024

ISBN: 979-8-3313-0662-5
ISSN: 1757-8981

Printed from e-media with permission by:

Curran Associates, Inc.
57 Morehouse Lane
Red Hook, NY 12571

Some format issues inherent in the e-media version may also appear in this print version.

Institute of Physics
Dirac House, Temple Back
Bristol BS1 6BE UK

Phone: 44 1 17 929 7481
Fax: 44 1 17 920 0979

techtracking@iop.org

Additional copies of this publication are available from:

Curran Associates, Inc.
57 Morehouse Lane
Red Hook, NY 12571 USA
Phone: 845-758-0400
Fax: 845-758-2633
Email: curran@proceedings.com
Web: www.proceedings.com

TABLE OF CONTENTS

IOP Conf. Series: Materials Science and Engineering **1315** (2024) 011001 doi:10.1088/1757-899X/1315/1/011001

Preface: 13th Ranshofener Leichtmetalltage

What strategies are available to minimize resource consumption in the light metal industry through recycling and energy efficiency? How can lightweight design make mobility more sustainable? And what role will artificial intelligence play in future processes in the metal processing industry? These and other questions are addressed at the 13th Ranshofener Leichtmetalltage 2024, which take place on the 26th–27th of September 2024 at the Hotel Gut Brandlhof in Saalfelden, Austria, and are organized in accordance with the criteria of the Austrian Ecolabel for Green Events.

Under the guiding theme of "Light Metals Innovations for Environmental and Economic Sustainability", this year's conference offers an exciting program in three sessions "Digitalization in the Context of Circularity", "Sustainable Process Development" and "Innovative Light Metals and their Characterization". The focus is on decarbonization and digitalization in process and material development as well as the material characterization of light metals. A balanced spectrum of international presentations from universities, RTOs and industry provides participants with an up-to-date overview of the latest scientific findings and successful light metal applications.

The 13th Ranshofener Leichtmetalltage 2024 are thus clearly focused on decarbonization and digitalization – even in times of diverse challenges. The first session, "Digitalization in the Context of Circularity", highlights best-practice examples from the aluminium industry, the balance between sustainability and performance in metal processing and the implementation of complex geometries using wire-based additive manufacturing.

The second session, "Sustainable Process Development", is dedicated to the decarbonization and reduced material consumption of industrial processes, for example through GigaCasting, or the use of innovative recycling technologies.

The third session, "Innovative Light Metals and their Characterization", addresses forward-looking topics such as the development of special aluminium alloys for additive manufacturing, the use of old car scrap for innovative alloys and the decarbonization of the magnesium industry. The conference proceedings contain scientific contributions to the conference. We would like to thank all the authors for their high-quality work and the participants of the 13th Ranshofener Leichtmetalltage 2024 for their interest. We wish you stimulating reading and creative impulses!

Scientific committee
Dr. Christian Chimani[1,2]
Dr. Thomas Klein[1]
Dr. Johannes Österreicher[1]
Prof. Dr. Peter Uggowitzer[3]

Organizing committee
Dr. Christian Chimani[1,2]
Mag. Florian Hainz, BA[2]
Denise Häuslmann[1]
Andrea Höfner[2]

[1] LKR Light Metals Technologies, AIT Austrian Institute of Technology, Lamprechtshausenerstr. 61, 5282 Ranshofen, Austria
[2] Center for Transport Technologies, AIT Austrian Institute of Technology, Giefinggasse 4, 1210 Vienna Austria
[3] Dept. Metallurgy, Chair of Nonferrous Metallurgy, Montanuniversität Leoben, Franz-Josef-Str. 18, 8700 Leoben, Austria

IOP Conf. Series: Materials Science and Engineering **1315** (2024) 011002 doi:10.1088/1757-899X/1315/1/011002

Peer Review Statement

All papers published in this volume have been reviewed through processes administered by the Editors. Reviews were conducted by expert referees to the professional and scientific standards expected of a proceedings journal published by IOP Publishing.

- **Type of peer review:** Single Anonymous
- **Conference submission management system:** Morressier
- **Number of submissions received:** 13
- **Number of submissions sent for review:** 12
- **Number of submissions accepted:** 8
- **Acceptance Rate (Submissions Accepted / Submissions Received × 100):** 61.5
- **Average number of reviews per paper:** 1.09
- **Total number of reviewers involved:** 10
- **Contact person for queries:**
  Name: Thomas Klein
  Email: thomas.klein@ait.ac.at
  Affiliation: LKR Light Metals Technologies, AIT Austrian Institute of Technology

IOP Conf. Series: Materials Science and Engineering **1315** (2024) 012001 doi:10.1088/1757-899X/1315/1/012001

Data Models for Casting Processes – Performances, Validations and Challenges

A M Horr[1*], R Gómez Vázquez[1] and D Blacher[1]

[1] AIT Austrian Institute of Technology GmbH, Center for Transport Technologies, Light Metals Technologies Ranshofen, Austria

*E-mail: Amir.Horr@ait.ac.at

Abstract. Data-driven models with their associated data learning and training schemes can be utilised for the light metal casting processes. This paper presents the basis of data model building processes along with data training and learning exercises for vertical direct chill casting and high pressure die casting (HPDC) applications. The concepts of efficient database building, data translations and sampling, as well as real-time model building and validations are briefly discussed. Rigorous performance studies were additionally carried out for two real-world case studies. Different combinations of data solvers and interpolators are adapted for the model building techniques, while machine learning schemes are used for data trainings.

1. Introduction

The data-driven and hybrid physical-data driven techniques are transforming the process modelling schemes toward more digitally oriented and cyber-based manufacturing processes. These digitalisation transformations promote more energy efficiency, productivity and innovation through faster and more effective modelling. The casting process is one of the most traditional material processes, which has been used to manufacture metallic and plastic parts. For the case of light metals like aluminium and magnesium, various casting processes are used to produce quality parts with reasonable strength [1]. The data real-time and reduced models can play a significant role in the digitalisation of these casting processes by increasing the pace of initial design and controlling of casting machines [2]. This would reduce the downtime preparations, and costly iterative trial and errors, while saving material and cost for new process designs [3]. There are various opportunities to improve the material processes like continuous direct chill casting, for instance, during planning of a new casting set up (e.g., alloy component) and/or choosing suitable set of parameters. Even though numerical simulations can help to optimise these processes, they are computationally expensive to evaluate the outcome for a single set of process parameters [4].

This paper presents an overview over the creation of efficient data models, their performances, validations, and challenges for casting processes. The role of appropriate combinations of data solvers and interpolators along with effects of data learning and trainings were examined and issues of applicability and reliability have been scrutinized. Two real-world case studies for vertical direct chill casting (VDCC) and multi-cycled high pressure die casting (HPDC) have been considered. The potentials for predictions and estimations of responses based on variations of process input parameters have been examined for further process optimisations and controlling.

2. Casting Processes – Modelling and Validations

Different types of casting processing such as die-casting, low pressure casting, HPDC, and electromagnetic die casting have been used in manufacturing industries to produce quality parts and billets. Hence, various modelling techniques have similarly been developed to model these multi-physical and multi-scale processes, including analytical, numerical, and hybrid physical data-driven techniques. By increasing the computational power in recent decades, numerical simulations of casting processes become more popular, where sophisticated and multi-physical finite element (FE) and computational fluid dynamics (CFD) techniques have been utilised for process modelling.

2.1 Modelling of Continuous Direct Chill Casting Process

The modelling and simulation of die chill casting process including its cooling and solidification sub-processes involve multi-physical and multi-phase considerations. Additionally, to improve the microstructure and mechanical properties of cast parts, the multi-scale simulations are sometimes employed using micro-macro interaction concepts for modelling of grain evolutions during the casting processes. These casting processes can be designed in vertical, horizontal or curved orientations where appropriate cooling systems are provided to produce pre-designed shaped billets [5,6]. Figure 1 shows the schematic casting processes for continuous and semi-continuous vertical, horizontal, and curved setup.

Some of the most popular numerical simulation techniques for the continuous casting processes are the fluid-thermal, coupled fluid-mechanical and mechanical simulations. These

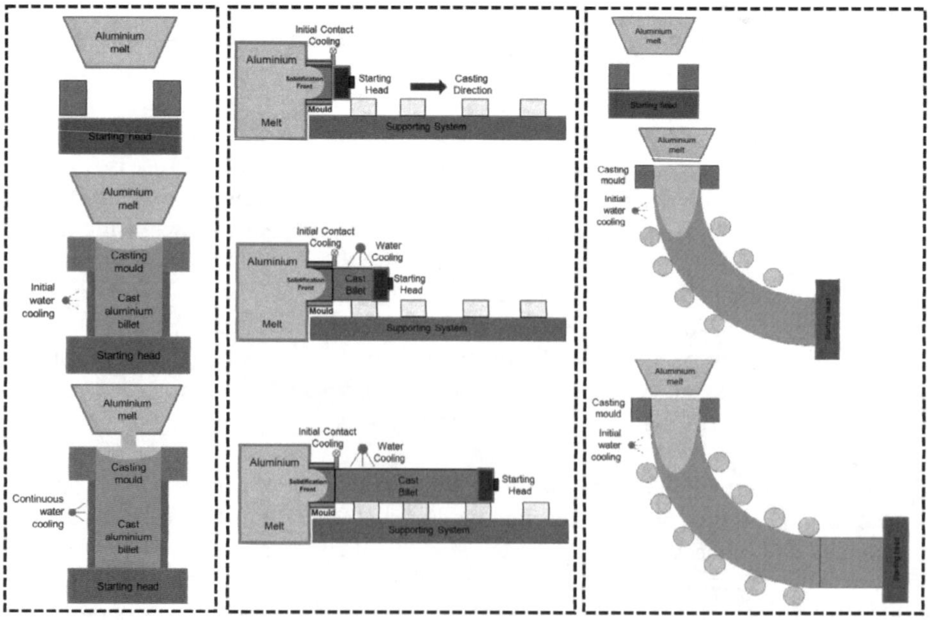

(a) (b) (c)

Figure 1. schematic views of; a) vertical; b) horizontal and c) curved die chill casting processes

simulations can be performed as a single phase or multi-phase scheme to calculate the melt dynamics, thermal energy evolutions, solidification and stress-strain state. For the solidification modelling the macro or micro species equations can be solved to determine the solidification at melt-solid interface.

2.2 Modelling of HPDC Process

HPDC processes are types of casting processes where pressurised melt materials are injected into the die to produce complex-shaped parts, typically in successive cycles. The high input pressure can generate a melt jet at nozzles, in which jet breakup phenomena might occur inside the die to fill cavities. Depending on the pressure, different melt flow regimes can be formed inside the die such as Rayleigh, first wind-induced, second wind-induced, and even atomization regimes. Most of the HPDC processes for light metal casting industries operate at low to medium Reynolds numbers (i.e., not at high atomized pressures) where Eulerian, Lagrangian, and hybrid Eulerian-Lagrangian modelling techniques can be employed. Figure 2 shows the HPDC machine at LKR laboratory with its typical CFD simulations.

3. Database Building – Scenarios for Simulations

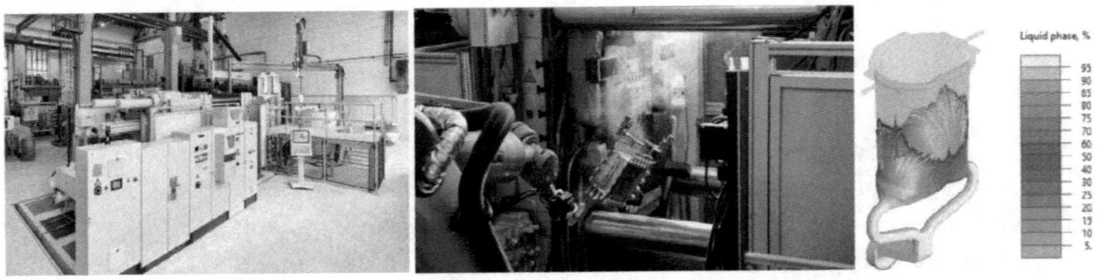

Figure 2. HPDC machine at LKR laboratory with its typical numerical simulation

Data models can accurately perform only when they are build based on appropriate size databases with right data density. Process data can come from different live or off-line sources including live sensor data and/or off-line experimental studies and numerical simulations. For the digital shadowing framework in this study, only off-line verified simulation results are used to generate minimal size databases for data model buildings. The following steps are pursued to achieve an appropriate database:

- Main predictive and corrective objectives of the real-time data models for casting processes were identified and their scopes are defined.
- Most influential process parameters with their practical ranges are defined and a multi-dimensional search space was established.
- Process database structure was designed and number of basic and extra design of experiments (DOEs) scenarios were outlined.
- A snapshot matrix for simulation scenarios was defined and variation of process parameters within the search space were considered.
- Simulation results were filtered and collected into the mapping module to create structured data at equidistance intervals.
- Process database was populated using mapped data for further data solver-interpolator procedures.

Process database building procedures usually involve handling of large amount of data and extensive filtering and mapping are mostly required to build a structured sematic database for the casting processes. The efficient sampling and balanced representation of the multi-dimensional search space have a profound effect on the accuracy and generality of the data model predictions.

4. Data Solvers for Casting applications

Different data techniques including data decomposition, projection, regression, and clustering schemes are used to disentangle the complex data pattern for dynamic processes. These data methods are essential for developing process real-time models including dynamic processes like vertical die chill and HPDC. Singular value decomposition (SVD) method is a very popular eigenvalue-based data solver which was developed as a general form of proper orthogonal decomposition method (POD). For dynamic processes with varying process parameters, snapshot matrix S that represent the multi-dimensional search space can be created using appropriate sampling technique (e.g., Sobol, LHS). The covariance matrix for the time- dependent responses of casting processes can then be calculated as;

$$[C(x,t)] = \frac{1}{N} SS^T \qquad [C(x,t)] = [R].[X(x,t)] \tag{1}$$

where $[\boldsymbol{R}]$ is the estimated system characteristics (using data techniques) and $[\boldsymbol{X}(\boldsymbol{x},\boldsymbol{t})]$ is the spatial and temporal input values (e.g., using experimental and simulation results). To decompose the data related to the complex responses of the processing system with varying input parameters, eigenvalue decomposition can be performed as;

$$[\boldsymbol{C}(\boldsymbol{x},\boldsymbol{t})] = \boldsymbol{U} \sum \boldsymbol{V}^{\boldsymbol{T}} \tag{2}$$

where U represents the spatial eigenvector decomposition, $\sum$ is the singular values diagonal matrix (e.g., eigenvalues) and $\boldsymbol{V}^{\boldsymbol{T}}$represent temporal decomposition for the transient casting process responses. To obtain the projection of snapshot matrix on most significant modes, only modes with higher energy in $\sum$ can be used to reconstruct the processing system. This will significantly reduce the order of the system equations and enable the real-time predictions for running processes. Alternatively, the principal component analysis (PCA) technique which is the statistical version of SVD may be employed to decompose the casting process responses. In this technique, the response data are projected onto the principal components as;

$$[\overline{\boldsymbol{C}}(\boldsymbol{x},\boldsymbol{t})] = \boldsymbol{V}^{\boldsymbol{T}}\boldsymbol{S} \tag{3}$$

where $[\overline{\boldsymbol{C}}(\boldsymbol{x},\boldsymbol{t})]$ is the transformed response in the principal components space. Hence, the data variance is preserved in this technique while the order of system equations is reduced without losing the major system characteristics. This simplifies the problem by reducing the dimensionality of the data, making the data model more computationally efficient while retaining essential features.

5. Case Study I: Continuous Direct Chill Casting

Motivation: The continuous casting setup benefits from use of data models: By exploring various process parameters and variations of initial water-cooling systems, the temperature homogeneity could be improved. This would reduce the risk of tearing (e.g., cold/hot cracking) and avoid scrap. There are several challenges to establish a suitable advisory system to help optimise continuous direct chill casting. To create a digital twin or shadow of the casting process, concepts of real-time predictions and corrections should be integrated into the virtual framework. For the initial system design and proof of concept, the feasibility study of data models for the vertical casting process has significant advantages over the horizontal case:

- The physical model needs to be calibrated to properly reflect the heat transfer of the casting process. To adjust the parameters for an accurate prediction of the thermal field, temperature measurements from within the solidifying billet are required. The exact positioning of thermocouples and handling of accompanying cables in the horizontal casting machine often poses a challenge. On the contrary, the vertical casting process with an open top allows for exact and relatively easy insertion of thermocouples.
- Numerical CFD models of the horizontal casting need to consider buoyancy effects due to gravity, thus the computational mesh needs to span from the top to the bottom of the billet. While one half could be arguably mirrored with suitable symmetric boundary conditions, this still poses a large computational effort. In the vertical case, on the other hand, the geometry can be further reduced to quarter model or even to a quasi-2D symmetric wedge (round billets formats), eventually allowing for shorter simulation times.

Numerical setup: A fluid-thermal CFD simulation framework has been set up to study the thermal evolution and solidification phenomena during the vertical DC casting process. The open-source solver directChillFoam [7] has been here employed, as an extension on top of OpenFOAM simulation package [8]. The validation case from directChillFoam project [7], based on the vertical DC casting experiment by Vreeman et al. [9], was taken as reference. A general description of the case setup is publicly available in the official documentation [10]. Only minor changes have been applied to the original case, mainly for allowing the parametrical study required to train\validate the data model with different combinations of initial melt temperature, cooling flow rate and casting speed (see Table 1).

A fixed domain strategy has been utilised, where the entire mesh for the numerical domain has been created at the start of the simulation. Material properties correspond to the binary alloy Al-6wt% Cu. The phase change obeys the temperature-dependent tabular values of the melt fraction (e.g., from a CALPHAD routine) and the solute properties are set to follow the Lever-rule. The heat transfer coefficients (HTCs) at the mould-melt interface (primary cooling) were locally averaged, depending on the value of the solid fraction. The heat transfer at the billet-water interface (secondary cooling) was introduced as tabular values obtained from the correlation proposed in [4].

All simulations were run in transient mode to ensure a correct description of the physical behaviour at steady process conditions. The runtime was set to 2,000s, more than doubling the period required for reaching quasi-steady state (see figure 3). Time-averaged results over the last 1,000s of temperature and melt-fraction magnitudes were saved for all cell-centers and then exported in CSV format, along with their spatial information (i.e., X- and Z-coordinates) for allowing further postprocessing (i.e. data re-ordering).

IOP Conf. Series: Materials Science and Engineering **1315** (2024) 012001 doi:10.1088/1757-899X/1315/1/012001

Table 1. Snapshot matrix for the vertical direct-chill casting case. Base scenarios are designated by "scenario B#", while validation scenarios are designated by "scenario DOE#".

Scenario	Initial melt temperature [K]	Water flow rate [l/min]	Casting speed [mm/s]
B1	943	72	1.2
B2	973	36	0.7
B3	988	108	2.5
B4	923	150	2
B5	938	90	1
B6	915	54	1.8
B7	993	180	0.5
B8	908	60	3
B9	893	168	2.8
DOE1	903	81	1.4
DOE2	933	99	1.9
DOE3	968	141	0.85

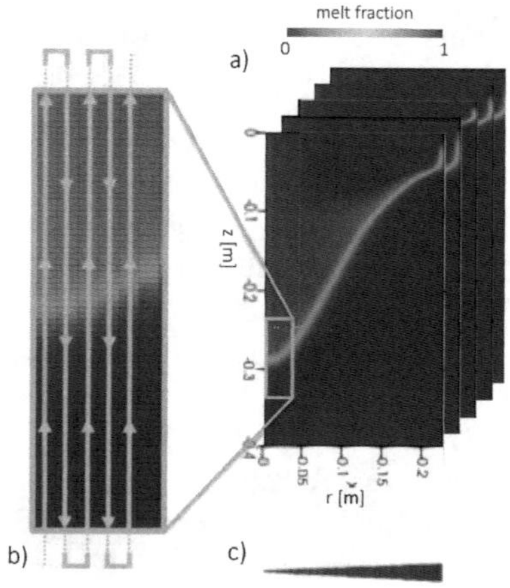

Figure 3. Basic contour results used to generate main database; a) 2D contours from simulation scenarios; b) Data processing strategy to minimize sudden thermal discontinuities between rows; c) Side view of wedge domain.

6. Case Study II: High Pressure Die Casting (HPDC)

Motivation: Within the scope of high-pressure die casting of aluminium alloys, finding optimal process parameters is also a challenging task, traditionally relying on long experienced personnel along with numerous *"trial-and-error"* iterations. Despite of the availability of numerical CFD models that allow simulating the intricate interactions among the numerous physical phenomena involved, the required long calculation times tend to limit the application. Hence, the number of process scenarios that can be studied using CFD are limited to a few representative cases. In contrast to physics-based models, trained data models may deliver similar results in a timely manner, thus allowing to envisage otherwise unfeasible tasks such as a complete mapping of the parameter field or even real-time process predictions based on current parametrical deviations.
Numerical setup: The goal of the data models was to reproduce the thermal evolution in different regions of the die over the course of several cycles. Hence, a series of scenarios were prepared based on a simplified initial setup. The simulations were run using the simulation software NovaFlow&Solid [11]. This commercial CFD package, especially aimed for industrial solidification processes, allows for easy re-utilization of case templates for parametrical studies.

The numerical domain accounted for all individual die parts, cooling channels, ingate system and chamber. The simulations were set to run 6 complete cycles, including air-blowing and water-spray chilling steps in between. The casting material was AlSi9Cu3(Fe) cast aluminium and the die material, steel (AISI H13). The initial temperature of all die parts was initially set to 220°C, as well as the temperature of the shot chamber. The temperature of the cooling channels was set to 80°C (water coolant) and to values between 120-160°C (oil as coolant), respectively. The piston was assumed at 100°C. The piston curve was defined in the initial setup, so that only its max. speed varied from one scenario to another. The other variable selected was the initial melt temperature (when being poured into the chamber). The parameter snapshot matrix used for training/validating the data model is shown in Table 2. Figure 4 shows typical CFD results for the multi-cycle temperature time-history of a HPDC process (with six filling cycles).

Table 2. Snapshot matrix for the high pressure die casting case. Base scenarios are designated by "scenario B#", while validation scenarios are designated by "scenario DOE#".

Scenario	Initial melt temperature [°C]	Max. piston speed [m/s]
B1	690	2.7
B2	675	3.3
B3	665	2.3
B4	645	2.9
B5	650	3.7
B6	680	3.9
DOE1	640	3.1
DOE2	670	2.6

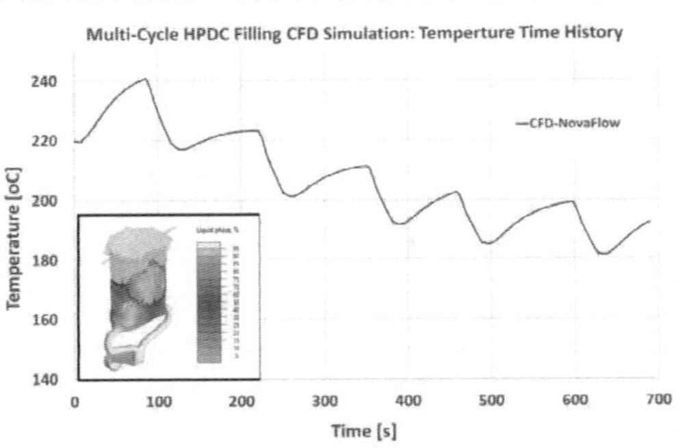

Figure 4. Typical CFD results for multi-cycle temperature time-history for HPDC process (six filling cycles)

The data provided to the data models were sampled at 6 locations, arbitrarily distributed over the different die elements, along the XZ-plane (Y-axis aligned with the piston displacement; Z-axis aligned with the vertical).

7. Discussion

The performance, validation and accuracy of data real-time models for direct-chill casting processes with their multi-phase and multi-physical aspects need to be evaluated carefully and practical challenges should be given attention for industrial applications. In this research work, the eigen-based SVD, regression and clustering solvers along with Kriging, radial base function (RBF) and inverse distance (InvD) data interpolators has been adapted to build efficient and accurate data models. The data training can also be performed using a single or double layer error back propagation neural network scheme with RBF function.

Figure 5 compares the temperature from the CFD to the SVD-Kriging data models for the vertical direct-chill casting case in a section of the computational mesh. The oscillating patterns

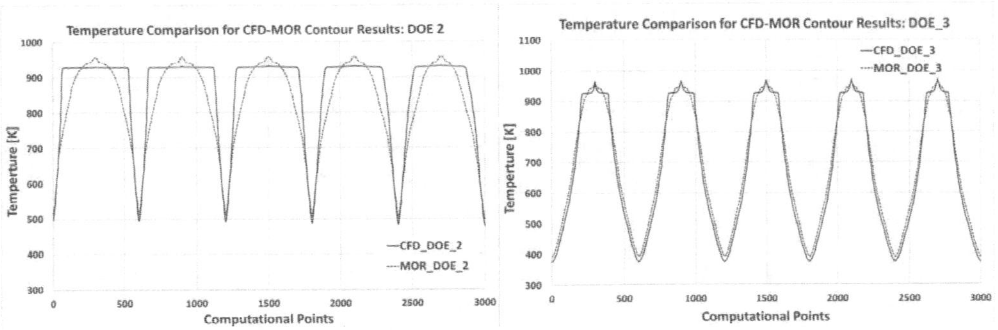

(a) **(b)**

Figure 5. Temperature results for computational points along contour lines; a) DOE2 and; b) DOE3.

stem from the data sampling path which covers the entire mesh in a snake like fashion (see figure 3b for schematic of the data processing path): starting with the central line of the billet going from the bottom (r=0, z=0.4m) towards the inlet (r=0, z=0m), the temperature rises; then down along the adjacent line to the bottom the temperature falls again. Figure 5 only shows the first 10 lines starting in the center (bottom-to-inlet and inlet-to-bottom making up one peak). The full data set covers the entire witdh from center to the billet surface. Each peak includes a sharp spike in the center which corresponds to the fixed inlet temperature boundary condition. To generate the predictions with the SVD-Kriging model (depicted as dashed line in figure 5), the following procedure was performed: Firstly, each CFD base result was mapped from the 2D mesh structure into a single vector following the described snake pattern. Then, the base scenario database comprised of one vector for each scenario B1-B9 (see table 1) was decomposed by the data solver. Finally, the data interpolator generated a new vector prediction for each process parameter combination given by the DOE cases.

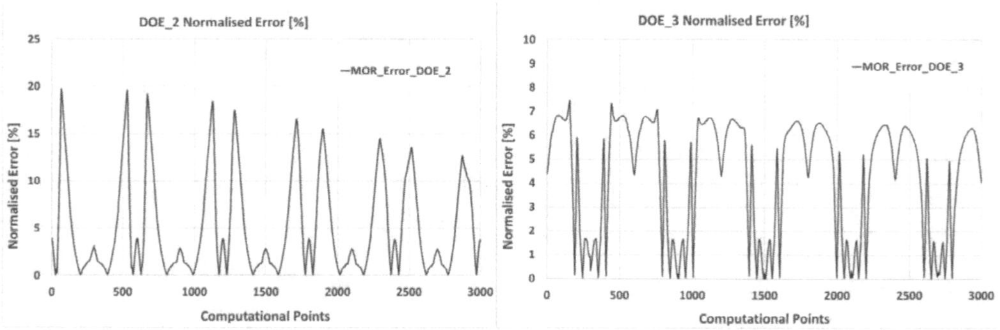

(b) **(b)**

Figure 6. Normalised error for computational points along contour lines; a) DOE2 and b) DOE3.

At the first glance, it is observed that the formulated data models can reliably be employed to produce real-time predictions of the temperature. Figure 6 shows the normalised error percentage for the predicted DOE scenarios at computaional nodes using the SVD-kriging data solver-interpolator for the same application. As it appears from these results, the reliability and accuracy of these data models are varying with the rate of data changes which in this case is the temperature gradient: at the bottom of the melt pool (which features a large thermal gradient near the solidification front), the prediction accuracy is decreasing. These challenges in the generation and training of process data models need to be overcome before they can reliably be employed within the digital twin concepts. These challenges can be listed as;

- For casting processes with their inherite multi-phase and multi-physical properties, data models need to be able to cope with rapidly changing data.
- The database size and data sampling within process parameters search space can considerably affect the prediction power of these models.
- To validation regimes need to be rigorous enough to consider normal, near boundary and extreme condition (extrapolation) scenarios for the process search space.
- Further data training and learning can also be performed using the Neural Network and/or other machine learing techniques to increase the accuracy and reliablity of data model for better predictive power.

For the HPDC case study, similar model building excersises have been carried out using different combination of data solver-interpolators. The multi-cycle nature of these processes makes the CFD numerical simulations long and tedious. Hence, very limited number of CFD simulations have been performed to build the database and the feasibilty study were undertaken

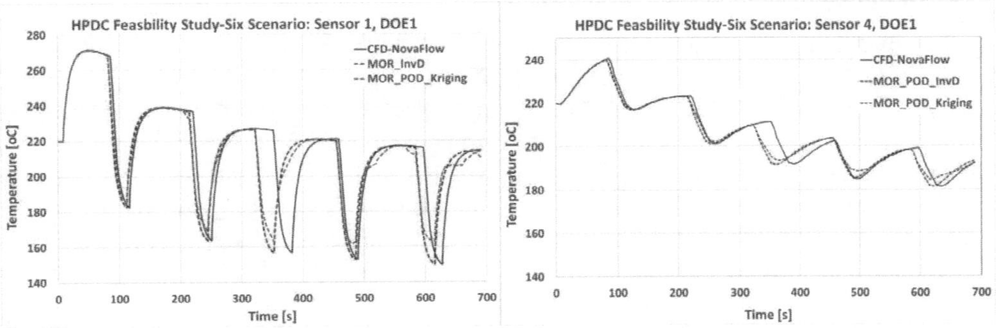

(a) **(b)**

Figure 7. Time-history temperature results for CFD and data models at; a) sensor 1 and; b) sensor 4

using limited number of scenarios. Figure 7 shows the estimation of time history cyclic temperature for two of the sensors compared to CFD simulations using SVD-InvD and SVD-Kriging data models. The computational time for the CFD simulation is about 54000 seconds for the six filling cyclic HPDC (i.e., wall clock time), while it takes only about 1.8 second for the data models to estimate the time history responses. As it appears from the results, with only six basic scenarios (a relatively small database), data models were strugling to predict the solidification time during each cycle. Hence, they have lagged behind the next filling cycle of the die, resulting in a time shift

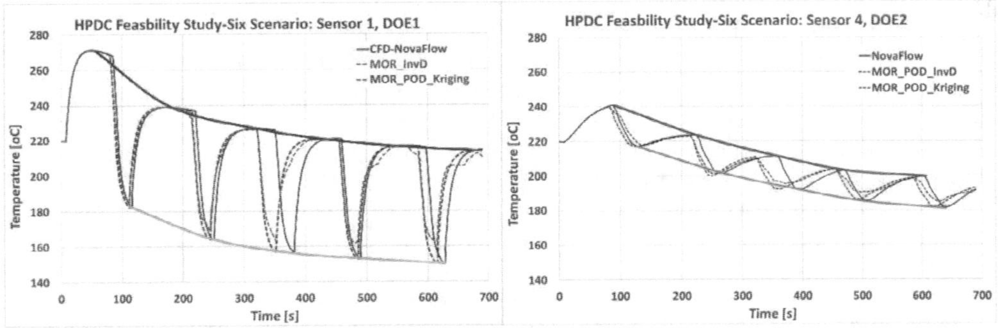

(a) **(b)**

Figure 8. Time-history temperature results with overlying descending temperature curves; a) sensor 1, DOE 1 and; b) sensor 4, DOE 2.

for the whole predictions. However, even with the small size database, data models provided relatively good predicions of the maximum and minimum temperature values of each cycle despite the shift in time. Figure 8 shows the CFD and data model predictions with overlaying decending temperature curves (maximum and minimum) for the six filling cycles. These are important informations for casting operators, since the overall temperature trend is an important parameter for the production effiecency in HPDC. While more scenarios are needed to fill a bigger and richer database for efficient HPDC data models, this feasbility study shows a lot of potential for utilising these models.

8. Concluding Remarks

The utilisation of data models for material process modelling can enhance the optimisation of existing processes and accelerate the design of future processes. In this paper, the performance, validation, and challenges related to these fast data models have been scrutinised to evaluate their accuracy and performances for material casting processes. In the first part of the manuscript, a brief representation of the direct chill casting and HPDC processes and their modelling and simulation techniques were illustrated, and some technical characteristics of these processes were discussed. The basic concept of casting processes' data models and their associated database building schemes were presented next, and the efficient use of data solver-interpolator combinations were discussed. In the following parts of the paper, two industrial case studies for VDCC and HPDC multi-cycle filling processes have been explained and their technical modelling challenges were examined. The performance, validations, and accuracy of process data models were scrutinised at the final part of manuscript, where major challenges related to the utilisation of these fast models were discussed.

From the presented results, it can be observed that proper combination of data solvers and interpolators can produce satisfactory results. However, the issues of high heating\cooling rates during casting processes can to some degree be problematic for the data models. Nevertheless, with further data training and learning, it is possible to achieve better results even for the processes with high gradient heating\cooling rates. The results of more detailed investigations on the best performance "solver-interpolator" combinations will be topic of a further publication.

Acknowledgment

The technical/financial support by the Austrian Federal Ministry for Climate Action, Environment, Energy, Mobility, Innovation and Technology, by the Federal State of Upper Austria and by the Austrian Institute of Technology (AIT) are gratefully appreciated. The financial support by Austrian Research Promotion Agency (FFG) for the *opt1mus* project (FFG-Nr. 899054) and by the European Commission for the *metaFacturing* project (HORIZON-CL4-2022-RESILIENCE-01, PID 101091635) are also greatly appreciated.

References

[1] Luo, A. A. (2021). Recent advances in light metals and manufacturing for automotive applications. CIM Journal, 12(3), 79–87. https://doi.org/10.1080/19236026.2021.1947088.

[2] Horr A. M. (2021). Optimization of Manufacturing Processes Using ML-Assisted Hybrid Technique, Journal of Manufacturing Letter, https://doi.org/10.1016/j.mfglet.2021.10.001.

[3] Horr A. M. (2020). Notes on New Physical & Hybrid Modelling Trends for Material Process Simulations, Journal of Physics: Conference Series, 1603(1), https://doi.org/10.1088/1742-6596/1603/1/012008.

[4] Weckman D.C., Niessen P. (1982). A numerical simulation of the D.C. continuous casting process including nucleate boiling heat transfer. Metall Trans B 13, 593–602. https://doi.org/10.1007/BF02650017.

[5] Horr A. M, Kronsteiner J. (2021). Dynamic material process simulations: Hybrid-evolving technique, Journal of Metals, 11(12), 1884; https://doi.org/10.3390/met11121884.

[6] Horr, A.M.; Kronsteiner, J. (2020). On Numerical Simulation of Casting in New Foundries: Dynamic Process Simulations. Metals, 10, 886. https://doi.org/10.3390/met10070886.

IOP Conf. Series: Materials Science and Engineering **1315** (2024) 012001 doi:10.1088/1757-899X/1315/1/012001

[7] Lebon B. (2023). directChillFoam: an OpenFOAM application for direct-chill casting. Journal of Open Source Software, 8(82), 4871, https://doi.org/10.21105/joss.04871.
[8] OpenFOAM-10 (2022). The OpenFOAM Foundation: https://openfoam.org.
[9] Vreeman C.J., Schloz J.D. & Krane M.J.M. (2002). Direct Chill Casting of Aluminium Alloys: Modelling and Experiments on Industrial Scale Ingots, *J. Heat Transfer* **124** 947-953. https://doi.org/10.1115/1.1482089.
[10] directChillFoam Documentation. (2023). https://blebon.com/directChillFoam.
[11] NovaFlow&Solid (2022). NovaCast Systems AB. https://www.novacast.se/product/novaflowsolid/.

IOP Conf. Series: Materials Science and Engineering **1315** (2024) 012002 doi:10.1088/1757-899X/1315/1/012002

From CAD model to WAM component: implementation of complex geometries using the example of a hydrogen tank.

Mayrhofer Florian[1*], Motlik Oliver[2], Pragerstorfer Dominic[2], Ucsnik Stephan[1], Schneider-Bröskamp Christian[1]

[1] LKR Light Metals Technologies, AIT Austrian Institute of Technology, 5282 Ranshofen, Austria.
[2] PEAK TECHNOLOGY GMBH, 4615 Holzhausen, Austria

*E-mail: florian.mayrhofer@ait.ac.at

Abstract. Hydrogen is expected to play a decisive role on the way to climate neutral transportation. This paper explores the innovative use of wire arc Directed Energy Deposition (waDED) in the manufacturing of a double walled, vacuum insulated liquid hydrogen storage tank. The study demonstrates the feasibility of the waDED process for the manufacturing of tank shells and furthermore the integration of heat exchanger channels into tank shells. Key findings include the development of a suitable process chain with integration of multiple features to ease manufacturing. However, challenges such as thermal shrinkage of the shells as well as geometrical deviations of the outer tank heat exchanger are encountered. Future research will focus on improvements in the process reliability with regards to geometric accuracy as well as in-depth testing under liquid hydrogen condition.

1. Introduction

Since hydrogen is one of the key technologies to achieve the climate neutrality and zero pollution economy by 2050 the hydrogen storage is a crucial field of research that requires advancements (1). In the mobility sector, most current systems store gaseous hydrogen at high pressures, up to 700 bar, necessitating stringent safety standards (2,3). To fulfil the safety requirements and withstand the high pressure, tanks are often manufactured from stainless steel or carbon fibre overwrapped metal versions (4). Due to the weight of stainless-steel tanks the gravimetric storage density is comparably low (2-5 kWh/kg) (5,6). To be more suitable especially for the aviation industry storage of liquid hydrogen tanks manufactured in lightweight design would increase the gravimetric storage density significantly (5).

The wire arc Directed Energy Deposition (waDED) process is especially suitable for thin, single wall and conformal shaped structures (7). Therefore, the technology is already established in the manufacturing of rocket components (8,9). With this work we aim to demonstrate the capabilities of the waDED process in the manufacturing of liquid hydrogen storage tanks. By integrating heat exchanger components into the tank shells using the waDED process, we further demonstrate the additive manufacturing approach, reducing weight and space requirements, and thereby increasing the gravimetric storage density of the system. Such innovations are essential for the future of sustainable aviation. We outline the entire process chain from the initial Computer Aided Design (CAD) model to the finalized tank components, to provide comprehensive insights into the use and complexities of this advanced manufacturing process.

2. Methods and experimental setup

While the main design work on the CAD geometry for the design of the double wall vacuum + multi layer insulation (MLI) insulated storage system made from aluminium (Figure 1) was conducted in CATIA, minor design modifications to improve the waDED manufacturability were conducted using Mastercam 2023 ® in which the Laser-aided-manufacturing module (LAM) by ModuleWorks GmbH is embedded. The LAM module was used to generate all tool paths for the trials and the manufacturing of the heat exchangers and the tank shells. The preparation of the toolpath into the sufficient robot motions were conducted in Robotmaster 2023 ®.

The welding setup used includes a KUKA KR 300 robot combined with a rotary and tilting table so angled torch positions can be welded in flat (PA) position. The rotary table contains an included tap water cooling and is capable of endless rotation. TransPuls Synergic 4000 CMT by Fronius International GmbH is used as power source for welding. Argon 5.0 (Messer Austria GmbH) was used as shielding gas at a flow rate of either 15 or 18 l/min depending on the wire diameter. The heat exchanger geometries were manufactured with ER 5183 wire with 1.2 mm diameter. For the tank shells the same alloy with 1.6 mm diameter was used. Both wires used were in form of a 140 kg drum.

Preparation of the base plates and intermediate machining was conducted on an ENCOMAT turning machine and an EMCO-Famup MC 75-50 CNC mill. The final machining of the tank shells was conducted by partner Peak Technology using a Hermle c52 u mt CNC mill. The 3D-scanning was performed using a Hexagon Absolute Arm with seven axis.

3. waDED of liquid hydrogen storage tank with integrated heat exchanger: Feasibility demonstration

The printed and machined tank as well as a cross section through the CAD geometry of the LH2 tank is displayed in Figure 1 to improve the understandability of the manufacturing steps. The dimensions of the storage tank are approximately 750 mm outer diameter and 700 mm length without the dome. The storage was manufactured in 4 separate shells to ensure the manufacturing, machining, and assembly process. The wall thickness of the inner and outer tank shell was estimated and confirmed with a finite element simulation (FEM) The inner tank shell has a wall thickness of 3 mm to withstand an operating pressure of 10 bar of liquid hydrogen. Since the outer tank shell is just subject to the external atmospheric pressure due to the internal vacuum, its wall thickness was further reduced to 1.9 mm. To reduce the system volume and thermal losses, the heat exchanger to extract gaseous hydrogen from the tank is integrated into the tank shell. The storage volume of the inner tank is approximately 150 l which results in a storage capacity of 10.5 kg of liquid hydrogen.

For the manufacturing of the geometry by waDED multiple difficulties arise:

- Thin tank shell and complex heat exchanger with required tightness for Hydrogen.
- Integration of complex heat exchanger channels and connectors.
- Tight tolerances compared to part size.

IOP Conf. Series: Materials Science and Engineering **1315** (2024) 012002 doi:10.1088/1757-899X/1315/1/012002

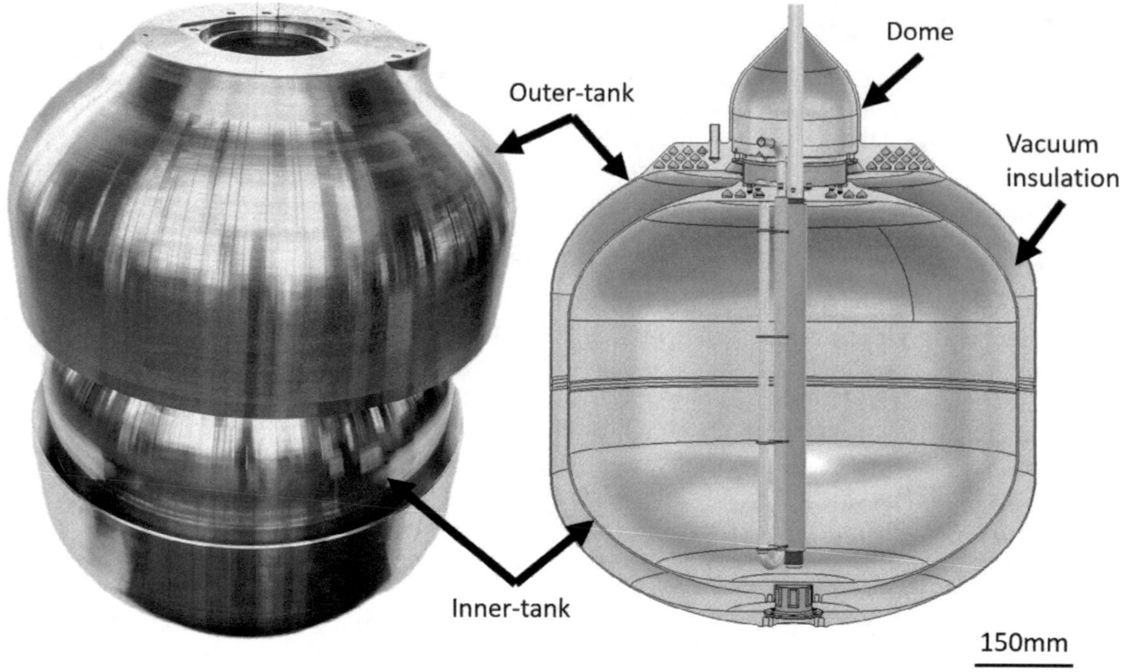

Figure 1. Partly assembled waDED manufactured LH2 storage tank with integrated heat exchanger. The left image shows the partly assembled tank with the LH2 tank on the inside and the vacuum insulation tank on the outside (opened for visualization). The cross-section of the CAD-model on the right illustrates the integrated heat exchanger channels within the tank shells.

Four main sections can be defined to outline the process steps for manufacturing. These sections, despite the assembly, will also serve as the titles for the following chapters:

- Iterative development of CAD-model including manufacturing strategy
- Development and preparation of manufacturing
- Manufacturing process with iterative machining
- Assembly of tank components

The definition of each section and it's containing tasks are visualized in the flowchart in Figure 2. While the flowchart appears sequential, it's essential to consider all subsequent processes in every predecessor decision. This approach supports manufacturability and ensures simplicity and efficiency.

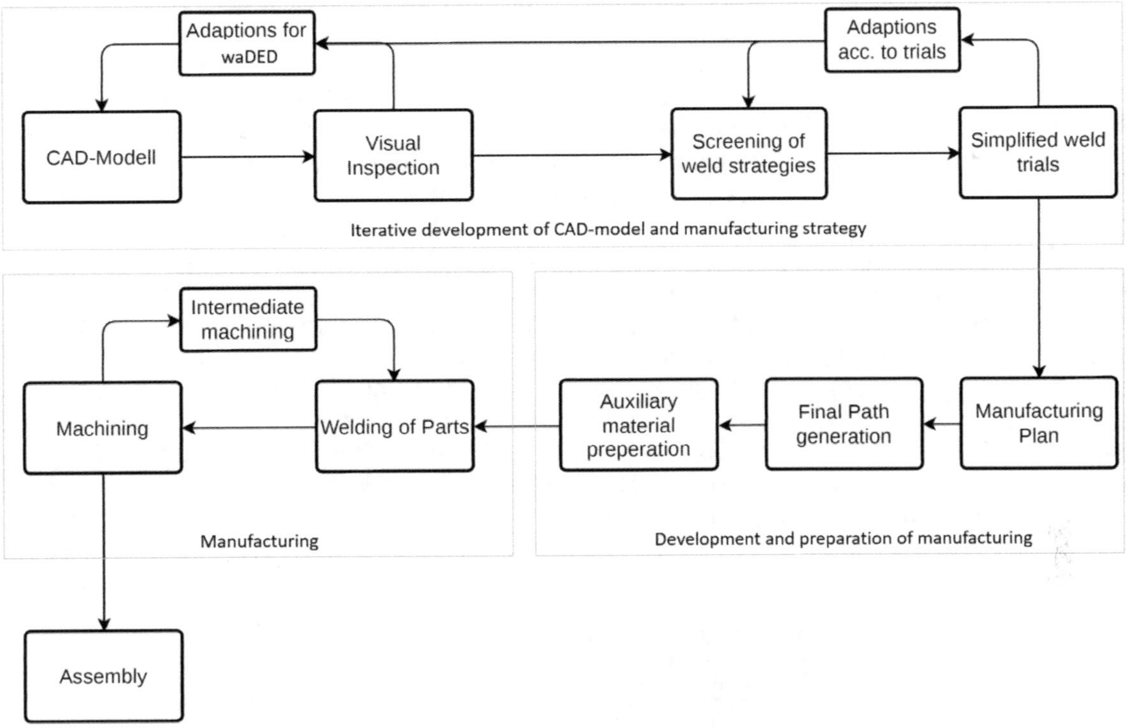

Figure 2. Flowchart of the process steps to develop and pursue a manufacturing strategy for the creation of complex waDED Geometries. The approach involves intermediate and final machining steps.

3.1 Iterative development of CAD-model and manufacturing strategy

Since the waDED process often comes to use to improve existing geometries or assess the process capabilities, CAD-model drafts are often already available. For complex geometries such drafts are required to go into detailed planning considering also a waDED manufacturing strategy. Figure 3a displays a draft model of the LH2 tank to be printed. The heat exchanger design is preliminary and visualizes the pipe routes and all the interconnections between the channels. The blue tubes carry hydrogen, while the separate red tubes can be flushed with water to inject heat.

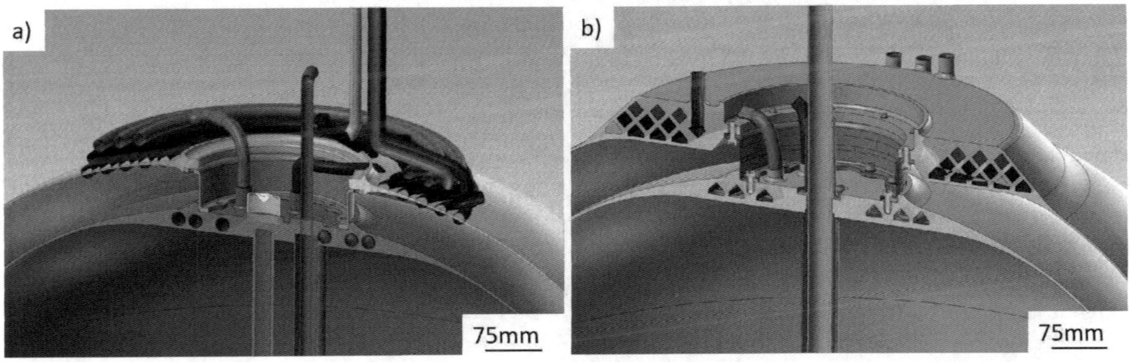

Figure 3. a) Cross-section of model draft with functional integration of the heat exchanger pipes and the connecting elements. b) Cross-section of the final model with integrated heat exchanger into the tank shells. Optimizations according to welding trials implemented in the design.

A deep understanding of the waDED manufacturing process is essential, to build upon the model. Simplified trials can be conducted to support critical manufacturing decisions. For instance, the build orientation of the tank shells was a profound decision. Figure 4a, b show robot process simulations. In Figure 4a, the tank shell is manufactured like a dome, starting at the largest diameter. In Figure 4b, the shell is manufactured like a bowl, starting at the centre (clamping materials not displayed). Each orientation has multiple pros and cons regarding base plate costs, weld path complexity, cooling capabilities, and more.

The main concern with the dome approach was closing the shell towards the centre. A trial was conducted to close off the top of a cylinder, to test this complexity and its feasibility. Figure 4c shows the robot simulation, where a cranked torch is used to weld inside the cylinder. Initially, the torch is perpendicular to the base surface. As the print progresses and the diameter reduces, the torch gradually tilts out of the cylinder to a maximum angle of 45°. With the angled torch it is critical to have no deviations between the programmed and welded layer height. Otherwise, the weld seam shifts sideways. Figure 4d shows the continuously welded result of the trial. The weld is uniform and flat almost up to the centre. The blue-marked area is critical due to rotational speed limitations of the robot axis. Although further optimization could resolve this issue, the risk of such complex welding strategies at the end of a final part is too high. Therefore, the tank shells were manufactured using the bowl-shaped approach (see Figure 4b).

Figure 4. Assessment of different welding orientations to manufacture the tank shells. Path simulation of orientation a) starting at the tank seam on a build plate. The shell would be manufactured like a dome. In orientation b) the tank shell would be manufactured like a bowl starting at a stock cylinder. c) Weld path simulation of simplified trial to verify if the dome approach is doable by trying to close of a cylinder. d) Welding result showing good results until centre of the disc. Axis speed limits result in height deviations (circled in blue).

Similar trials were conducted to assess methods to manufacture the ambos integrated heat exchanger for the inner tank shell. These results could also largely be applied to the outer heat exchanger, as the goal was to maximize common features, to simplify production. Previous Figure 3b shows the cross section of the final tank design. The main adaptions were conducted on the heat exchangers and were:

- Flattening of the heat exchanger shape
- Adapting channel cross section to stackable diamond pattern
- Accessibility to all heat exchanger channels straight from the outside

3.2 Development and preparation of manufacturing

As stated earlier, the whole manufacturing chain is considered from early stage on. Therefore, a general concept for the manufacturing was already conducted in the CAD model design phase. Nevertheless, the finalization was made with the adapted model and in close collaboration with the machinists. The tank shells were manufactured starting on EN-AW 5083 cylinders with 80 mm diameter. The cylinder was clamped concentric on the rotational welding table using a specially designed and manufactured clamp which included a water flow to enable cooling or heating of the cylinders bottom. Tank shells without heat exchanger were welded sideways on the cylinder. The cylinders partly remained in the tanks even after machining. Additionally, the precisely machined cylinders were utilized as probing geometry for the subsequent machining of the tank. In terms of concentricity this was sufficient, while the exact longitudinal placement was figured with a 3D scan of the tank shells to compensate geometrical deviations resulting from the print process.

The tank shells containing the heat exchanger were printed onto a 20 mm thick EN-AW 5083 baseplate. The baseplates partly remained in the final tank and therefore served as tight barrier between the inner tank volume and the heat exchanger. Since the heat exchangers are difficult to manufacture, in case of welding defects this measure ensures tightness of the tank. The lower half of the first channel was machined into the plates to reduce the amount of printed material and to further utilize the plates. Additionally, a machined centre pocket and side groves allowed for exact placement on the groove table of the waDED setup, while also serving as probing geometry on the mill.

An overview on the process steps for the manufacturing of the heat exchangers including their shells is shown in Table 1. From this overview, two particularly effective practices can be highlighted. Firstly, the implementation of clamping sections into the design for waDED. During manufacturing of the heat exchanger of the inner tank a ring was printed on top, which was machined and used for clamping in a three-jaw-chuck for the welding of the tank shell onto it. Additionally, the machined ring served as probing geometry for machining since the outer shape of the base plate, which previously defined concentricity, had to be removed to print the tank shell on. Secondly with the same tank shell, the previously printed heat exchanger channels were utilized for preheating and cooling of the part during the shell printing. The channel inlet and outlet were machined, threaded and hose connectors were attached. Especially during the print this method of internal cooling proved very effective.

Table 1. Overview on process steps for fabrication of tank shells containing the heat exchanger components. Similarities in the manufacturing process are visualized between inner and outer tank shell.

	Inner Tank	Outer Tank
Baseplate	- 20mm thick baseplate to reduce deformation - Half cooling channel milled in base plate - Milled pockets to index centre and x-axis orientation 	- 20mm thick baseplate to reduce deformation - Half cooling channel milled in base plate - Milled pockets to index centre and x-axis orientation - Two borings for central clamping 
Heat exchanger	- Welding of heat exchanger channel - Weld filling between heat exchanger channel - Welding of clamping ring - Welding of connectors to channels 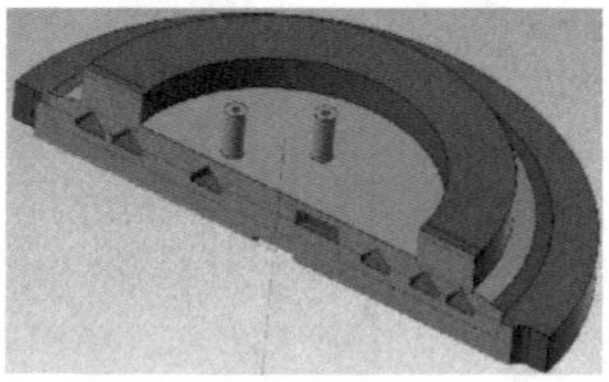	- Alternate welding of channel walls and filling layers - Connectors welded solid and opened during machining - Flat top surface without connectors to ease clamping during welding of the tank shell 
Intermediate machining	- Machine outside of clamp ring concentric to base plate - Transfer indexing pockets from base plate to clamp ring surface - Machine of access material of the base plate - Drill and thread connectors 	- Milling top surface of heat exchanger → Clamp surface for second operation - Transfer indexing pockets on the base - Machine of access material of the base plate 
Tank shell	- Clamp heat exchanger in three-jaw-chuck on clamp ring - Connect heat exchanger channels to cooling/heating device - Welding of tank shell 	- Clamping on groove table - Heat exchanger channels not accessible → use of table heating/cooling due to big contact surface - Welding of tank shell 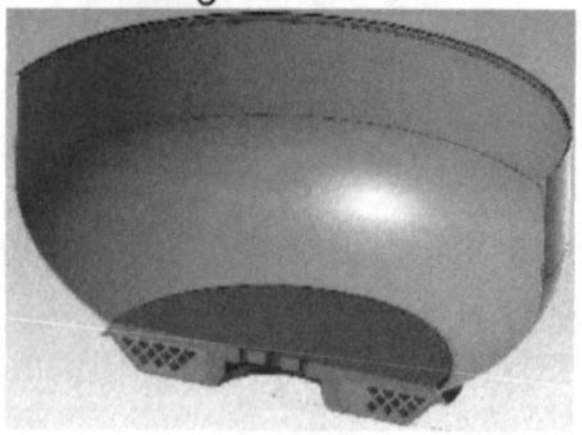

3.3 Manufacturing process with iterative machining

With the CAD geometry complete, weld paths finalized, weld parameters optimized from early-stage trials, and auxiliary materials prepared, the manufacturing of the tank shells commenced. Due to the novelty of manufacturing such geometries the order for the manufacturing of the shells was selected with ascending complexity. In the first stage, both tank shells without heat exchanger were printed. The path strategy combined a multi-track strategy in the dome area to achieve the required wall thickness and single track with spiral weaving in the rest of the tank shell. With each shell the parameter for the layer height could be optimized to reduce geometry deviations. The manufacturing steps of the heat exchanger shells were already depicted in Table 1.

The final machining of the shells was conducted by PEAK TECHNOLOGY GmbH and required in depth machining knowledge due to the large component size and the thin walls. Figure 5a shows a shell in the CNC milling machine after the first few passes but not with the final shape yet. In the dome section the waDED surface is still visible. Figure 5b depicts the fully machined outside of the shell with no defects visible. Additionally, the image shows the clamp ring added to all shells as last step of the waDED. The ring solely serves the purpose to ease the machining process. This highlights a big advantage of the additive manufacturing process. With little efforts supporting geometries for subsequent process steps can be implemented.

Figure 5. Machining process of the outside of an inner tank shell. The shell is screwed onto the machine table using a waDED manufactured clamp ring around the tank. a) Early machining stage with multiple unmachined sections on the surface. b) Tank shell with final outer dimension and completely machined surface finish.

3.4 Assessment of the feasibility of fabricating a double walled LH2 storage tank with integrated heat exchanger by waDED.

The waDED manufacturing of the four tank shells was successfully conducted. However, several challenges arose during manufacturing. Before printing the first tank shell, the thermal shrinkage of the shell diameter was uncertain. The unknown cooling rate made it difficult to accurately estimate the temperature equilibrium within the tank during printing. The weld path geometry was adapted, to compensate for shrinkage and geometrical deviations. The shells were

IOP Conf. Series: Materials Science and Engineering **1315** (2024) 012002 doi:10.1088/1757-899X/1315/1/012002

manufactured with an effective wall thickness of approximately 8 mm resulting in a deposition rate of approximately 2 kg/h. The diameter shrinkage during cooling off for the tank shells was measured post-cooling to be 2 to 4 mm. Using a process simulation could be beneficial in this context to optimize the path geometry and reduce excess material.

A structured approach, including an intermediate machining step, was crucial for the producibility of the heat exchangers. An early sample of the inner heat exchanger channel was tested for hydrogen tightness both at cryo- and room temperature. Therefore, the heat exchanger channels were attached to a hydrogen supply, the channel was flushed, and then pressurized with hydrogen at 6 bars (maximum available pressure). Tightness was checked using a hydrogen leak detector first at ambient, and after that with the heat exchanger submerged in liquid nitrogen. In both states, no hydrogen leakage could be detected.

The final assembly of the tank shell and the access components was not conducted due to time and budget limitations. Also, during final machining of the outer heat exchanger one channel was slightly milled into. This was probably caused due to a combination of multiple effects:

- Deformation of heat exchanger base plate during printing of approximately 2 mm
- Probing on CNC mill impeded due to deformation resulting in deviation of the machining position.
- Tolerances of the waDED process
- Thin outer wall thickness (3 mm) of heat exchanger

This could have been prevented by thickening the outer wall of the CAD-geometry or by further reducing or compensating the deformation. A repair using waDED after the fact is not possible, due to the thin wall thickness of the heat exchanger component. The heat input would melt and further open the channels, while having a lack of fusion in areas with thicker walls.

4. Conclusions

The proof of feasibility using waDED for manufacturing a LH2-WAM-Tank was largely successful and resulted in gaining knowledge in the manufacturing of medium sized components including complex structures. The main findings were:

- Tank shells with no heat exchanger could be manufactured with slight weld path adaptions to compensate for thermal shrinkage in diameter.
- waDED could successfully be used to support the subsequent milling steps by including clamp geometries.
- Manufacturing of inner tank heat exchanger was successful and showed no hydrogen leakage at room- and cryo-temperature.
- Outer tank shell heat exchanger was successfully manufactured, but it was not fully sealed after machining due to a partially open milling channel.

While the integration of channels into waDED-manufactured components has been explored in previous research(10,11), the heat exchanger design incorporated into the tank shell using waDED presented in this work is novel. We demonstrated the feasibility of manufacturing three stacked channels separated by a single printed wall. Minor improvements to the geometry could resolve the issue encountered during the machining of the heat exchanger. Future work will focus on the final joining and assembly of the tank shells. Additional areas for further research include optimizing the in-process control to reduce geometrical deviations as well as testing of waDED manufactured storage tanks with liquid hydrogen filling and thermal cycling trials.

IOP Conf. Series: Materials Science and Engineering **1315** (2024) 012002 doi:10.1088/1757-899X/1315/1/012002

5. References

1. European Commission. Communication from the Commission to the European Parliament, the Council, the European Economic and Social Committee and the Committee of the Regions. A hydrogen strategy for a climate-neutral Europe. [Internet]. 2020 [cited 2024 Jun 24]. Available from: https://eur-lex.europa.eu/legal-content/EN/TXT/?uri=CELEX:52020DC0301

2. Gómez JA;, Santos DMF, Salkuti R, Zhu D, Gómez JA, Santos DMF. The Status of On-Board Hydrogen Storage in Fuel Cell Electric Vehicles. Designs 2023, Vol 7, Page 97 [Internet]. 2023 Aug 2 [cited 2024 Jun 24];7(4):97. Available from: https://www.mdpi.com/2411-9660/7/4/97/htm

3. Rivard E, Trudeau M, Zaghib K. Hydrogen storage for mobility: A review. Vol. 12, Materials. MDPI AG; 2019.

4. Barthelemy H, Weber M, Barbier F. Hydrogen storage: Recent improvements and industrial perspectives. Int J Hydrogen Energy. 2017 Mar 16;42(11):7254–62.

5. Huete J, Nalianda D, Pilidis P. Impact of tank gravimetric efficiency on propulsion system integration for a first-generation hydrogen civil airliner. The Aeronautical Journal [Internet]. 2022 Aug 10 [cited 2024 Jun 25];126(1302):1324–32. Available from: https://www.cambridge.org/core/journals/aeronautical-journal/article/impact-of-tank-gravimetric-efficiency-on-propulsion-system-integration-for-a-firstgeneration-hydrogen-civil-airliner/CC45233F25B8A29C5916BFDD552D2056

6. Krainz G, Bartlok G, Bodner P, Casapicola P, Doeller C, Hofmeister F, et al. Development of Automotive Liquid Hydrogen Storage Systems. AIP Conf Proc [Internet]. 2004 Jun 23 [cited 2024 Jul 9];710(1):35–40. Available from: /aip/acp/article/710/1/35/1000381/Development-of-Automotive-Liquid-Hydrogen-Storage

7. Liu J, Xu Y, Ge Y, Hou Z, Chen S. Wire and arc additive manufacturing of metal components: a review of recent research developments. International Journal of Advanced Manufacturing Technology [Internet]. 2020 Nov 1 [cited 2024 Jun 25];111(1–2):149–98. Available from: https://link.springer.com/article/10.1007/s00170-020-05966-8

8. He T, Yu S, Runzhen Y, Bo Z. Oscillating wire arc additive manufacture of rocket motor bimetallic conical shell. International Journal of Advanced Manufacturing Technology [Internet]. 2022 Apr 1 [cited 2024 Jun 25];119(9–10):6601–12. Available from: https://link.springer.com/article/10.1007/s00170-021-08477-2

9. Gradl PR, Greene SE, Protz C, Bullard B, Buzzell J, Garcia C, et al. Additive manufacturing of liquid rocket engine combustion devices: A summary of process developments and hot-fire testing results. 2018 Joint Propulsion Conference [Internet]. 2018 [cited 2024 Jun 25]; Available from: https://arc.aiaa.org/doi/10.2514/6.2018-4625

IOP Conf. Series: Materials Science and Engineering **1315** (2024) 012002 doi:10.1088/1757-899X/1315/1/012002

10. Pragana JPM, Bragança IMF, Silva CMA, Martins PAF. Hybrid Wire-Arc Additive Manufacturing of Conformal Cooling Channels: A Feasibility Study. International Journal of Precision Engineering and Manufacturing - Green Technology. 2023 Jan 1;10(1):45–57.

11. Han S, Zhang Z, Ruan P, Cheng S, Xue D. Fabrication of circular cooling channels by cold metal transfer based wire and arc additive manufacturing. Proc Inst Mech Eng B J Eng Manuf [Internet]. 2021;235(11):1715–26. Available from: https://doi.org/10.1177/0954405421995613

Acknowledgments

The authors would like to thank the project partners TEST-FUCHS GmbH, TEST-FUCHS Aerospace Systems GmbH and Dr. Ryll-Lab GmbH for collaborative pursuit of this project.

CRediT authorship contribution statement

Florian Mayrhofer (Conceptualization; Methodology; Formal analysis; Investigation; Resources; Data curation; Writing – Original draft; Visualization) **Oliver Motlik** (Investigation; Writing – Review & Editing); **Dominic Pragerstorfer** (Investigation; Writing – Review & Editing); **Stephan Ucsnik** (Conceptualization; Writing – Review & Editing; Supervision; Project administration; Funding acquisition) **Christian Schneider-Bröskamp** (Conceptualization; Methodology; Investigation; Resources; Data curation; Writing – Review & Editing, Supervision; Project administration; Funding acquisition)

Declaration of competing interest

The authors declare that they have no known conflicts of interest that could have appeared to influence the work reported in this paper.

Data availability statement

The data underlying the present investigations can currently not be shared as it forms part of an ongoing study.

Funding

This research has been funded by the Austrian Federal Ministry for Climate Action, Environment, Energy, Mobility, Innovation and Technology within the framework “TAKE OFF” in the project “LH2-WAM-Tank” administered by FFG.

IOP Conf. Series: Materials Science and Engineering **1315** (2024) 012003 doi:10.1088/1757-899X/1315/1/012003

Insights on Weld Quality using Unsupervised Learning: Clustering of MIG/MAG Process Signals

Evans Doe Ocansey[1], Hannes Hinterbichler[2], Roxana-Maria Holom[1], and Michael Araz[3]

[1]RISC Software GmbH, Hagenberg im Mühlkreis, Austria
[2]Fronius International GmbH, Pettenbach, Austria
[3]BENTELER Automobiltechnik GmbH, Paderborn, Germany

E-mail: roxana.holom@risc-software.at

Abstract. With the increasing complexity of gas metal arc welding (GMAW) processes, data-driven approaches for monitoring and understanding automated GMAW production lines are gaining increased prominence. In this work, welding process data recorded in a production environment is analysed using unsupervised learning methods. We describe a data processing pipeline for feature engineering and apply a state-of-the-art clustering method to gain more insights into the welding production process. The clustering results are compared with the results from the application of dimensional reduction techniques and discussed based on human-interpretable characteristics of the welding process.

1 Introduction

Gas metal arc welding (GMAW) is a key production process in metalworking industries. Modern arc welding processes are highly dynamic, with a multitude of different process variants available [1]. In addition, driven by market demands and automation, the complexity of modern welding systems is rapidly increasing [2]. For human operators, monitoring and optimizing the welding production process is becoming ever more challenging. With the advancing digitization of our world, modern information technologies are also finding their way into welding production lines [3]. Emerging technologies like big data and artificial intelligence (AI) may provide solutions to support process monitoring, prediction and control, as well as quality inspection [2].

In this context, the European project called Data and Metadata for Advanced Digitalization of Manufacturing Industrial Lines (metaFacturing) has the overall goal of establishing a more resilient production process by researching the development of an AI-based digital twin for metal part production, thereby reducing costly rework and scrap resulting from out-of-specification parts [4]. Line operators will be assisted by AI-based applications that will provide enhanced information about the manufacturing process. Leveraging the integration of diverse types of data, such as welding process data, visual inspection data, and quality control data from destructive testing, within metaFacturing we are creating the data environment required to determine whether defects in weld seams correlate to information hidden in the characteristics of the corresponding data.

Currently, we are building a foundation for predicting the weld seam quality from the welding process data, i.e., defects can be identified earlier (maybe even online) in the process. This converges with the overall goal of significantly reducing rework, scrap, machine downtime and energy spent, hence increasing material and process efficiency. Therefore, in this paper, production data obtained from welding seams is used for the development of an unsupervised machine learning (ML) model. In the first step of analysing the raw welding signals, we apply a clustering algorithm. Several approaches that could prove to be effective for this task are available, like Density-Based Spatial Clustering of Applications with Noise (DBSCAN), Hierarchical Clustering or k-means clustering. In this research, we focus on analysing the raw signals of electrical welding current and voltage in the form of time series by applying the k-means clustering method. In a follow-up step, clustering algorithms will be combined with feature extraction from the chosen welding signals, to enable the detection of anomalies and the assignment of an anomaly score for each weld seam.

Many studies on approaches to unsupervised GMAW process monitoring have been reported in the literature. Sumesh et al. (2017) established a statistical correlation between welding process signals and welding defects by comparison of probability density distributions of welding current and voltage [5]. Kirchner et al. (2020) applied a long short-term memory (LSTM) autoencoder to detect welding process instabilities in multivariate welding process signals from the reconstruction error and the latent space of the autoencoder. The reported anomalies were largely short circuits prevented by the process control unit of the welding system [6]. Some studies combine welding process data with additional sensor data, e.g. with video data recorded during welding [7]. Reisch et al. (2020) trained ML models too in the context of wire arc additive manufacturing (WAAM). They combined the outputs of the different models processing welding process data and video data, to calculate an overall anomaly score employing a Mahalanobis distance [8]. Other studies do not consider the analysis of welding process data (like current and voltage), but rely on acoustic signals for monitoring and anomaly detection [9, 10].

In many of the studies mentioned above, the data is often generated in a laboratory setting, with relatively small sample sizes. The present studies focus on the application of clustering methods on data recorded in an industrial production environment. The remainder of the paper is organised as follows. In Section 2, we present an overview of the welding process, the motivation and the challenges of this work, and we introduce the production data used for the analysis. In Section 3, we detail the methodology for analysing the welding signals. In Section 4, we present the results and the feedback received from the welding (domain) experts. Section 5 summarizes the content of the paper and discusses ideas for future work.

2 Arc welding process (MIG/MAG)

2.1 Description and motivation

MIG- and MAG-welding are among the most common welding processes for steel and aluminium alloys. Both are jointly referred to as gas metal arc welding (GMAW) processes. GMAW is characterised by the electric arc which forms between the workpieces and the welding wire. This arc heats the joint while a protective gas (inert or active, thus metal inert gas (MIG) and metal active gas (MAG)) shields it from the environment. Using a robot, the welding torch is moved along the two workpieces' joints to create a weld seam with the desired length [1]. The quality of this joint is mainly given by the penetration of the created weld seam. It is a percentage of the workpiece thickness and a measure of the connection of both workpieces. A minimum penetration is an agreed quality feature between customers and suppliers of welded components to ensure the connection of both pieces for the expected life cycle.

For the determination of the penetration, destructive testing is broadly applied. Cutting a welded component perpendicular to the orientation of the weld seam gives an insight into the cross-section. The sample is then prepared by polishing and etching before the characteristics are measured visually. On the one hand, this procedure is fast and reliable, on the other hand, a part is destroyed and there is no data available other than for this sample.

A continuous control of the weld seam quality to estimate the penetration is our goal. This reduces the number of destroyed parts and can be applied to every single weld. Another benefit is early information on issues for the automatic or non-automatic optimisation of the process.

As a first step towards this goal, anomalous welds shall be isolated by the application of AI on the data from a welding power source.

2.2 Challenges

The weld penetration is dependent on many controlled and uncontrolled parameters. Considering the robot, the main parameters are the distance of the torch to the joint and the welding speed. Any changes in these parameters directly affect the arc-length, therefore the welding voltage, and the energy input per unit length of the weld seam. Since the process is dependent on the heat input from the arc, the robot thus has a significant influence on the quality [2].

The workpieces form a joint which is uneven along its length. The gap width, position, and contours vary between different parts and affect the heat distribution alongside the workpieces' heat treatment

Figure 1: (a) Illustration of the cross-section and (b) photo of the weld seam.

before welding.

The greatest impact on the weld quality can be expected from the welding power source. These electronic devices contain sensors and complex control functions to ensure the weld quality by automatic means. Due to the direct exposure to the process, wear-off of the mechanics and electrics needs to be monitored. The wire feed mechanics and contact tip in the torch have to be thoroughly maintained.

Due to the dependency of the welding parameters on one another, it is expected that most information is contained in the signals of the power source. The analysis of these signals to deliver an approach for the decryption of quality information will be described in the following chapters.

2.3 Data recording

The dataset utilized in this article was generated by a Fronius TPS/i welding system, installed within the production line for welding sessions of the project partner BENTELER Automobiltechnik GmbH, as part of the metaFacturing project. For further analysis, a particular weld seam was selected. It is schematically illustrated in Figure 1a, showing a butt joint of two aluminium profiles. As a welding process, Fronius Pulse Multi Control (PMC) is used. Moreover, the option Fronius Synchro Pulse is enabled, resulting in the characteristic ripple effect of the weld seam surface, see Figure 1b.

The welding process data of the weld seam was recorded for many different parts, resulting in 1478 recorded data samples. The description of the recorded welding process signals, as well as challenges regarding the merging of time series signals recorded at different sampling frequencies, are described elsewhere [11]. The main signals used in the present analysis are the voltage, `welding_voltage`, and the current, `welding_current`, of the arc during welding, sampled at 10 kHz. Moreover, signals used for communication between the welding system and the robot, like `arc_on` and `welding_start`, are used to segment the welding signals into different regions of interest (see Section 3.1). These signals were sampled at 1 kHz. The different sampling rates at which the welding signals were recorded pose a challenge since any proper analysis of the welding activity will essentially require that all signals be sampled at the same sampling rate. In [11], we describe these challenges in detail and how they were resolved. In particular, the dataset utilised within this article results from the upsampling of the low 1 kHz frequency signals to 10 kHz with linear interpolation to fill out missing data.

3 Analysis of welding signals

3.1 Data preparation

This section delineates the data preprocessing pipeline, as depicted in Figure 3. For each dataset corresponding to a weld seam of the specified machine serial number, the process initiates as follows: filter out the dataset based on the designated seam number to generate a pool of datasets per weld seam. In this case, this step yielded 1478 weld seam datasets. Subsequently, we clean the data by removing datasets containing potential outliers. Incomplete welding sessions and welding sessions where an error code was raised by the welding system, are excluded, resulting in 1459 weld seam datasets.

Next, the dataset undergoes further processing through either **Pipeline A** or **Pipeline B**, as illustrated in Figure 3. Within **Pipeline A**, the dataset is subjected to two distinct transformations. The first transformation selects the signals `welding_current` and `welding_voltage` for each dataset. The second transformation calculates the power spectral density (PSD) of the `welding_current` and `welding_voltage` using SciPy's Welch[1] function [12], producing two feature vectors per weld seam, each with dimensions $(1, n)$, where n is the number of frequencies. Merging the feature vectors sequentially by their respective production time results in 2 feature matrices, each with dimensions (m, n), where $m = 1459$, the number of weld seams. Finally, we merge these 2 feature matrices, which result in a tensor of dimensions $(m, n, 2)$.

[1] Documentation of SciPy `welch` function

Within **Pipeline B**, each dataset per weld seam is initially divided into three distinct phases using the low-frequency signals, `arc_on` and `welding_start`, which were upsampled with expert knowledge from Fronius. Figure 2 illustrates these phases, which are further described below:

- **Phase I** is the *start phase* at the beginning of the weld seam. In this phase, the arc is ignited, and the base material is heated.
- **Phase II** is the *main phase*, where welding is usually carried out at a constant welding speed.
- **Phase III** is the *end phase*, where the crater at the end of the weld seam is finished.

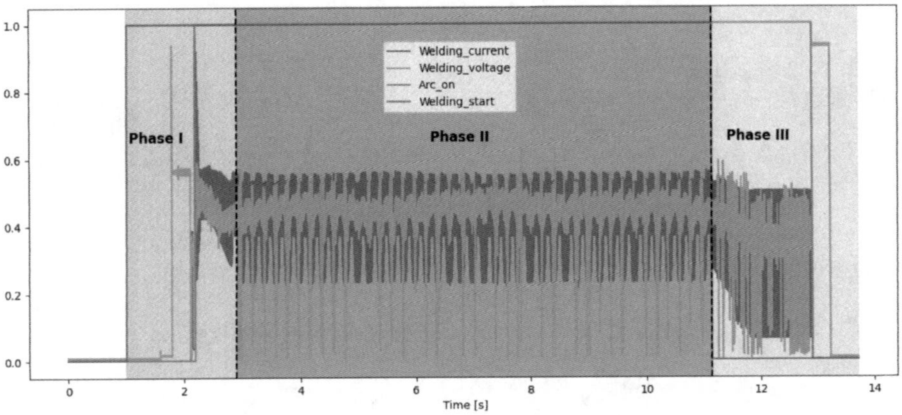

Figure 2: Separation of the welding data into three different phases. For illustration purposes, the depicted signals were normalized to the range between 0 and 1.

In this research work, we restrict the analysis activities that are described in the next subsection, to **Pipeline A** and Phase II (the main phase) of **Pipeline B**. As demonstrated in Figure 3 below, the Phase II dataset, generated for each weld seam, undergoes the exact transformation as described earlier for **Pipeline A** to obtain the feature tensor of dimension $(m, n, 2)$.

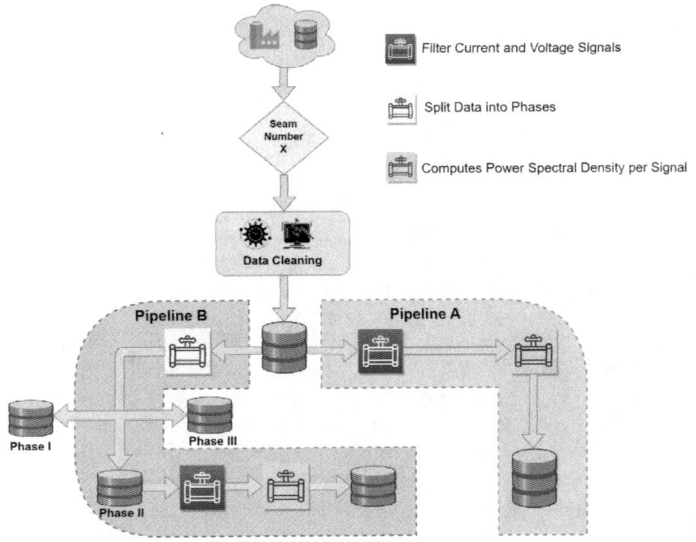

Figure 3: Data preprocessing and feature extraction pipeline.

IOP Conf. Series: Materials Science and Engineering **1315** (2024) 012003 doi:10.1088/1757-899X/1315/1/012003

3.2 Unsupervised learning – Clustering

This subsection details the machine learning methodology employed to analyse the dataset generated in Subsection 3.1. For an analysis focused on the quality statuses of the weld seams, it would necessitate the availability of target labels defining their quality, thereby indicating a supervised machine learning approach. However, due to the unavailability of quality data for the weld seams at the time of analysis, we adopted an unsupervised machine learning approach.

Specifically, we utilized the k-means clustering algorithm, as implemented in [13, 14], to cluster weld seams based on their PSD features. A primary challenge with the k-means algorithm is determining the optimal number of clusters. Generally, minimizing the cost function alone is not effective, as this approach tends to favour a larger number of clusters, given that increasing the number of clusters typically reduces the cost function. Hence, choosing the value of k solely to minimize the cost function is not a suitable method.

4 Results and discussion

Our approach to determining the number of clusters was experimental. We executed the k-means clustering algorithm, with the number of clusters ranging from 2 to 50. For each specified number of clusters, the algorithm was run with 50 random initialisations of the cluster centroids. The model with the minimum distortion cost for each chosen number of clusters was selected as the best model. Additionally, we computed the mean silhouette score for each chosen number of clusters, to evaluate the quality of clustering. The silhouette score measures how similar an object is to its own cluster compared to other clusters, ranging from -1 to 1, where a higher value indicates better clustering performance.

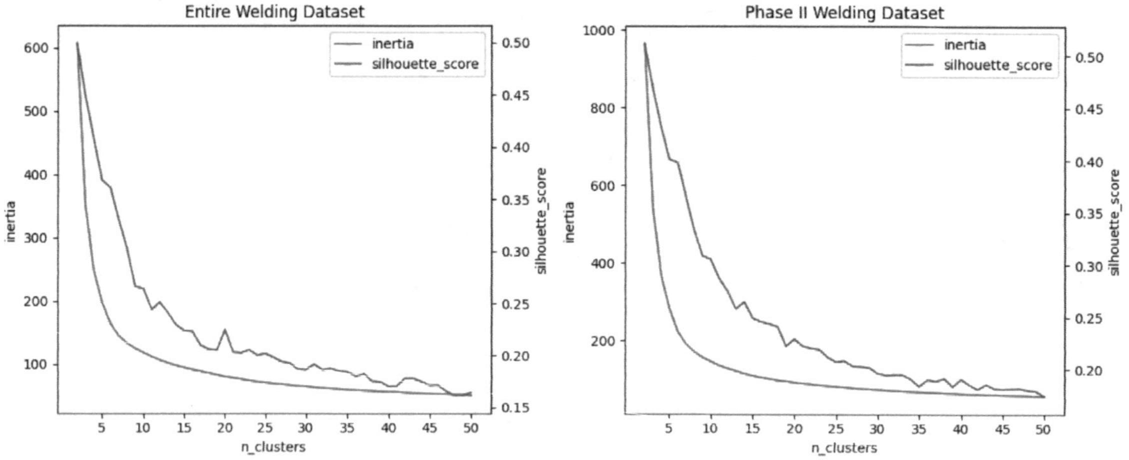

Figure 4: Number of clusters against inertia and silhouette score.

Figure 4 illustrates the experimental results for the datasets generated by **Pipeline A** over the entire duration of the welding session and by **Pipeline B**, which is confined to the *main welding phase* (Phase II). From Figure 4, it is evident that the silhouette scores indicate a decrease in clustering quality as the number of clusters increases. Consequently, we limited our analysis to 2 and 3 clusters.

For each specified number of clusters, we concatenated the feature matrices of the `welding_current` and `welding_voltage` from both **Pipeline A** and **Pipeline B**. We then performed Principal Component Analysis (PCA), reducing the dimensionality of the frequency feature space from 1025 to 2. The visualization of this dimensionality reduction, grouped by their respective clusters, is presented in Figure 5. For all shown cases, the two principal components together account for more than 95 % of the variance in the dataset, with frequencies ranging between 170 to 278 Hz as part of the top 10 frequencies having higher contribution scores to both components. Consequently, these frequencies contribute significantly to the variance captured within each of the principal components.

IOP Conf. Series: Materials Science and Engineering **1315** (2024) 012003 doi:10.1088/1757-899X/1315/1/012003

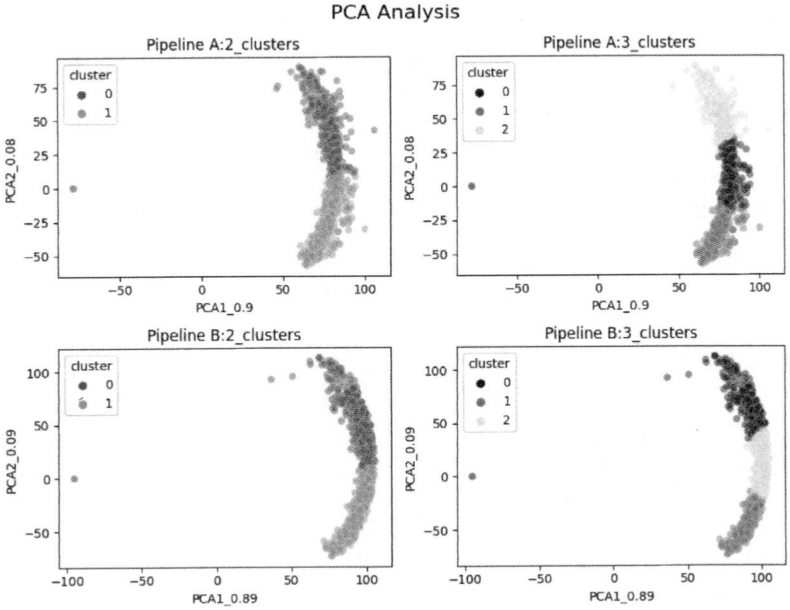

Figure 5: Downward projection of feature matrices using PCA and categorised based on their cluster values. The numeric values associated with the axis labels denote the explained variance ratio attributed to the respective principal component.

The cluster assignments predicted by the k-means clustering algorithm for each weld seam were used as target labels to train a classifier. The purpose of this classifier was to identify the frequencies that significantly influence the predicted clusters. Specifically, we employed the XGBoost classifier [15] to determine which frequencies are most critical in predicting the various clusters. The results, shown in Figure 6, highlight the top 10 frequencies for each of the k-clusters, with $k = 2, 3$, in the datasets generated by **Pipeline A** and **Pipeline B**.

The analysis reveals that, across all results, the most important frequencies fall within the range of 170 to 200 Hz. This finding is expected, as the welding pulse frequency, and therefore the frequency of droplet detachment, for the investigated weld seam lie within this range. Notably, for the welding current and the scenario involving 3 clusters (`psd_welding_current:3_clusters`), both pipelines indicate that the most important frequency is 366 Hz. This frequency likely represents a higher harmonic of the welding pulse frequency.

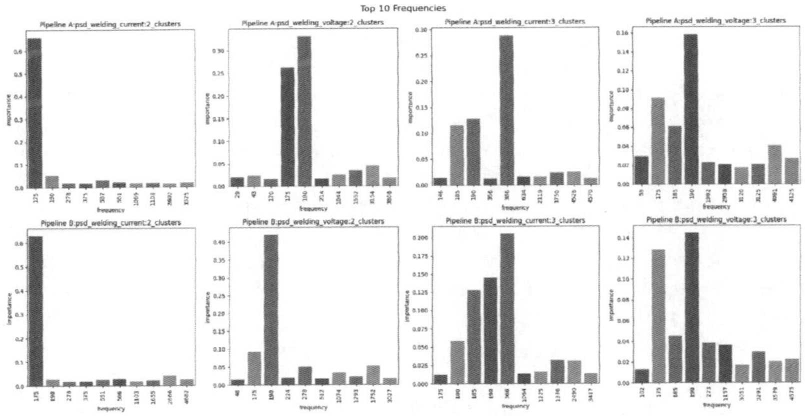

Figure 6: Top 10 frequencies obtained by a classifier predicting the cluster membership.

Figure 7 illustrates two scatter plots of the mean `welding_current` and mean `welding_voltage`, with each weld seam colour-coded according to its respective cluster. Here, we only focus on **Pipeline B** (scatter plots for **Pipeline A** are not shown). The mean current and voltage for each weld seam were calculated during active welding, specifically when the signal `arc_on` equals 1 (refer to Figure 2 above). It is evident that the differences between clusters are relatively well-demonstrated in the scatter plots in Figure 7. The clusters are primarily distinguished by varying values of welding current and voltage. This observation aligns with the PCA presented earlier (see Figure 5), where only two principal components account for the majority of the variance in the dataset, effectively separating the different clusters.

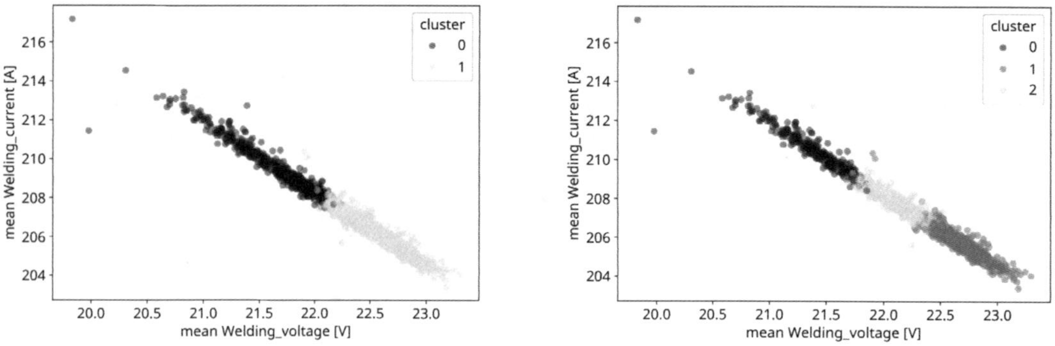

(a) (b)

Figure 7: Scatter plots of mean `welding_current` and mean `welding_voltage` for **Pipeline B**: Results for (a) two clusters and (b) three clusters.

5 Conclusion and future work

The analysis presented in this study demonstrates the application of unsupervised learning techniques, specifically k-means clustering, to identify patterns in welding process signals. The primary focus was on analysing the electrical signals of welding current and voltage recorded from a Fronius TPS/i welding system. The clustering results, based on power spectral density (PSD) features of these signals, revealed significant insights into the welding process quality and identified key frequency ranges that are most likely to influence weld seam quality.

Further analysis using principal component analysis (PCA) and feature importance analysis revealed that specific frequency bands, particularly those around the welding pulse frequency, in the range of 170 to 200 Hz, contribute significantly to the clustering results. The study's experimental approach to determining the optimal number of clusters provided valuable feedback from domain experts, indicating that a smaller number of clusters (two or three) is most likely to distinguish between different weld quality levels.

The successful application of unsupervised learning to industrial welding data in this study highlights the potential for such methods to improve process monitoring and quality control in real-world manufacturing environments. By identifying anomalies in weld seams, this approach can reduce the need for destructive testing, thereby saving time and resources while enhancing overall production efficiency.

Future work will focus on refining this approach by incorporating additional features and exploring different clustering algorithms. Furthermore, correlating the identified clusters with actual weld quality measurements will be crucial to validate the findings and establish a reliable anomaly detection system. Additionally, investigating the use of deep learning techniques for feature extraction and clustering could potentially improve the performance of the anomaly detection model. By continuously refining the approach and incorporating feedback from domain experts, we aim to develop a robust and effective tool for monitoring and improving welding process quality.

Acknowledgment

Funded by the European Union. Views and opinions expressed are however those of the author(s) only and do not necessarily reflect those of the European Union. Neither the European Union nor the granting authority can be held responsible for them.

The work described in this paper is supported by the metaFacturing project (GA 101091635), which has received funding under the Horizon Europe programme.

References

[1] Norrish J. Recent gas metal arc welding (GMAW) process developments: the implications related to international fabrication standards. Welding in the World. 2017;64:755-67.

[2] Wang B, Hu SJ, Sun L, Freiheit T. Intelligent welding system technologies: State-of-the-art review and perspectives. Journal of Manufacturing Systems. 2020;56:373-91.

[3] Norrish J. Evolution of Advanced Process Control in GMAW: Innovations, Implications, and Application. Welding Journal. 2024;103:161-75.

[4] metaFacturing Project Partners. Data and Metadata for Advanced Digitalization of Manufacturing Industrial Lines; 2023. Accessed: 2024-07-25. https://metafacturing.eu/.

[5] Sumesh A, Rameshkumar K, Raja A, Mohandas K, Santhakumari A, Shyambabu R. Establishing Correlation Between Current and Voltage Signatures of the Arc and Weld Defects in GMAW Process. Arabian Journal for Science and Engineering. 2017;42:4649-65.

[6] Kirchner V, Vishev I, Rossdeutscher M. Detecting process instabilities in industrial gas metal arc welding time series. Procedia Manufacturing. 2020;52:150-5.

[7] Bzymek A. Application of selected method of anomaly detection in signals acquired during welding process monitoring. International Journal of Materials and Product Technology. 2017;54(4):249-58.

[8] Reisch R, Hauser T, Lutz B, Pantano M, Kamps T, Knoll A. Distance-Based Multivariate Anomaly Detection in Wire Arc Additive Manufacturing. In: 2020 19th IEEE International Conference on Machine Learning and Applications (ICMLA); 2020. p. 659-64.

[9] Horvat J, Prezelj J, Polajnar I, Čudina M. Monitoring Gas Metal Arc Welding Process by Using Audible Sound Signal. Strojniški vestnik - Journal of Mechanical Engineering. 2011;57(3):267-78.

[10] Rohe M, Stoll BN, Hildebrand J, Reimann J, Bergmann JP. Detecting Process Anomalies in the GMAW Process by Acoustic Sensing with a Convolutional Neural Network (CNN) for Classification. Journal of Manufacturing and Materials Processing. 2021;5(4).

[11] Rodríguez-Echeverría, J and Ocansey, E D and Holom, R -M and Michno, T P and Hinterbichler, H and Meyer-Heye, P and Neves, J and Schnall, M and Hartmann, M and Gautama, S . Data Quality Strategies in Gas Metal Arc Welding Production for Machine Learning Applications; 2024. To be published.

[12] Virtanen P, Gommers R, Oliphant TE, Haberland M, Reddy T, Cournapeau D, et al. SciPy 1.0: Fundamental Algorithms for Scientific Computing in Python. Nature Methods. 2020;17:261-72.

[13] Tavenard R, Faouzi J, Vandewiele G, Divo F, Androz G, Holtz C, et al. Tslearn, A Machine Learning Toolkit for Time Series Data. Journal of Machine Learning Research. 2020;21(118):1-6. Available from: http://jmlr.org/papers/v21/20-091.html.

[14] Pedregosa F, Varoquaux G, Gramfort A, Michel V, Thirion B, Grisel O, et al. Scikit-learn: Machine Learning in Python. Journal of Machine Learning Research. 2011;12:2825-30.

[15] Chen T, Guestrin C. XGBoost: A Scalable Tree Boosting System. CoRR. 2016;abs/1603.02754. Available from: http://arxiv.org/abs/1603.02754.

Investigation of Wire-Arc Directed Energy Deposition of Copper for use in e-Mobility applications.

Kamalesh Bharadwaj[1*], Mathias Silmbroth[1] and David Mühlgrabner[2]

[1] LKR Light Metals Technologies, Austrian Institute of Technology, Ranshofen, Austria
[2] AVL List, Steyr, Austria

*E-mail: kamalesh.bharadwaj@ait.ac.at

Abstract. Additive manufacturing of pure copper is interesting for utilising high electrical and thermal conductivity properties for electrical as well as thermal applications. Wire-arc directed energy deposition is beneficial for producing large parts with less material wastage and is suitable for relatively less abundant resources like copper. The present study explores the feasibility and properties of parts produced using CuSn1 feedstock alloy and this process. Metallographic analysis and micro-hardness tests were conducted. The relative density of 99.99% is achieved, the micro-hardness is 64 $HV_{0.1}$ and the grain sizes are observed to be growing along the build direction.

1. Introduction

Wire arc directed energy deposition process (WA-DED, also known as Directed Energy Deposition-Arc and Wire Arc Additive Manufacturing) is an additive manufacturing (AM) process for building large components at low costs and minimal tooling. In this process a wire feedstock material is molten using energy from an electrical arc or plasma. Compared to powder-based AM, WA-DED parts have lower geometrical accuracy but achieve far higher material deposition rates.

Copper is an important metal due to its conductive properties and excellent recyclability. The demand for copper parts is increasing globally owing to the current focus on sustainable energy and transport solutions. The very high thermal and electrical conductivity, and reflectivity make the processing of pure copper difficult. The same properties are important for the industrial applications of copper. The differences in the properties of pure aluminium and copper are highlighted in **Table 1** [1,7,10].

Table 1. Comparison of properties of pure aluminium and copper [1,7,10].

Property	Aluminium	Copper
Density	2700 kg/m³	8910 kg/m³
Melting Temperature	660 °C	1083°C
Thermal Conductivity	230 W/mK	397 W/mK
Electrical Conductivity	37 MS/m or 65 %IACS	58.8 MS/m or 102 %IACS
Cross-section area for same conductivity	160 mm²	100 mm²
Weight for same conductivity	50 g	100 g
Tensile Strength	50-60 MPa	200-250 MPa
Elongation	40 %	60 %

The reflectivity of copper reduces the absorption of laser power and affects the processing with laser-based AM process [11]. However, this is not a challenge for WA-DED process. The effect of alloying elements on the electrical conductivity of copper is critical for selecting the alloy with a balance between properties and manufacturability. The effect of small percentage of alloying elements on the electrical conductivity is shown in **Figure 1** below.

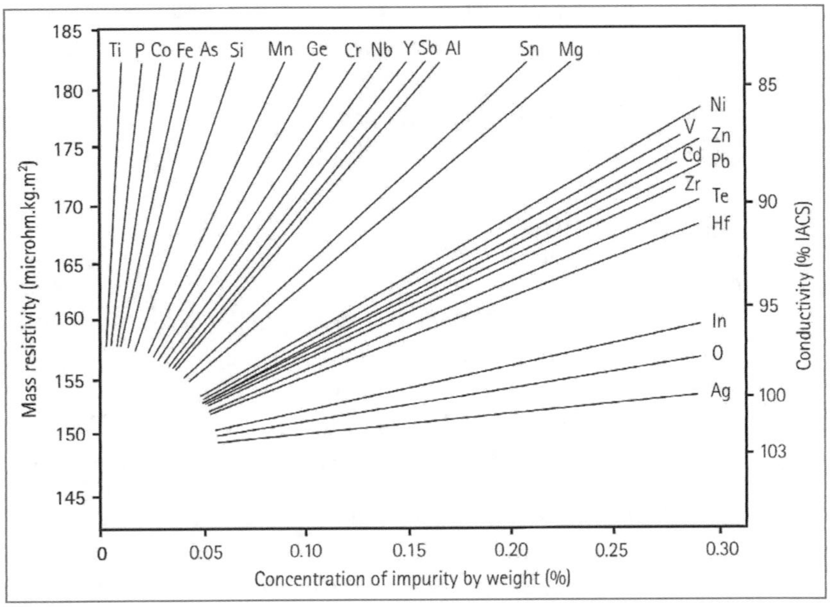

Figure 1. Effect of alloying elements on resistivity and conductivity of copper [7].

The processing of pure copper with wire-based processes has been studied by a handful of researchers. Baufeld studied the microstructure and properties of copper parts produced with wire electron beam AM (EBAM). Pure copper and CuSn1 were used to produce test samples. The manufacturability of copper using wire EBAM was verified, and deposition rates of 1.5 to 3.3 kg/h were recorded. Due to the vacuum environment in EBAM, very good surface quality and low oxidation levels were observed. Cracking in deposited walls was observed and assumed to be due to thermal stresses with scope of further investigations [2].

Williams et al. studied and compared the WA-DED processing of pure copper (electrical wire) and CuSn1 alloy (CW-1860) using gas tungsten arc welding process. The authors observed higher porosity for the pure copper and postulated that the faster cooling and solidification of the melt pool to be the cause of porosity formation in pure copper. The effect of local and global shielding (<600 ppm Oxygen) on both types of wire was studied and found to be significant for reducing the porosity levels for the pure copper wire. Low surface oxidation was observed for both the alloys with global shielding. They also observed that global shielding has no impact on the electrical conductivity of the walls. Electrical conductivity (IACS%) values of 20% for the CW-1860 alloy and 102% for pure copper were observed [3].

Baby, J., Amirthalingam studied the WA-DED deposition of CuSi3Mn1 alloy and compared the pulsed and short-circuited welding modes. They found that the short-cicuit process has contact with melt pool and produces equiaxed grains relative to columnar grains in pulsed mode and the parts have consistent tensile results [4].

The WA-DED of pure copper was studied by Deshmukh et al. by manufacturing walls from 99% pure copper alloy. It was found that the microstructure had coarse equiaxed grains along the deposition direction and epitaxial grain growth along the build direction. The mechanical properties observed were average micro-hardness of 88 $HV_{0.98N}$, UTS and elongation values of 226 MPa & 124.4 % along deposition direction and 219 MPa & 99.6 % along build direction [5].

In this study, the feasibility of WA-DED process using CuSn1 alloy and the properties in the produced parts are investigated. The process parameters are developed and evaluated on the basis of process stability, porosity levels and the degree of fusion between layers. The microstructure is analysed for understanding the grain structure and discovering anomalies if any. The developed process is evaluated by fabricating an electrical busbar demonstration part.

2. Materials and Equipment

2.1 Setup and Software

The study is focused on WA-DED using Cold Metal Transfer (CMT) arc welding process. A Fronius iWave CMT power source with a push-pull wire feeder system is used for gas metal arc welding. The torch is positioned in 3D space using a Kuka robot (6+2 axis) equiped with Arctech Advanced welding package. Preheating is carried out using an underneath positioned electrical heating block. Machine, sensor and process data are collected using an Iba data acquisition system.

Mastercam with ModuleWorks laser-aided manufacturing module and Robotmaster software are used for CAM toolpath and robot offline programming respectively. Metallographic preparation and imaging is carried out using calibrated Olympus digital microscope. Micro-hardness is evaluated using a Zwick DuraScan machine. The WA-DED setup and the offline programming environment in Robotmaster software are shown in **Figure 2**.

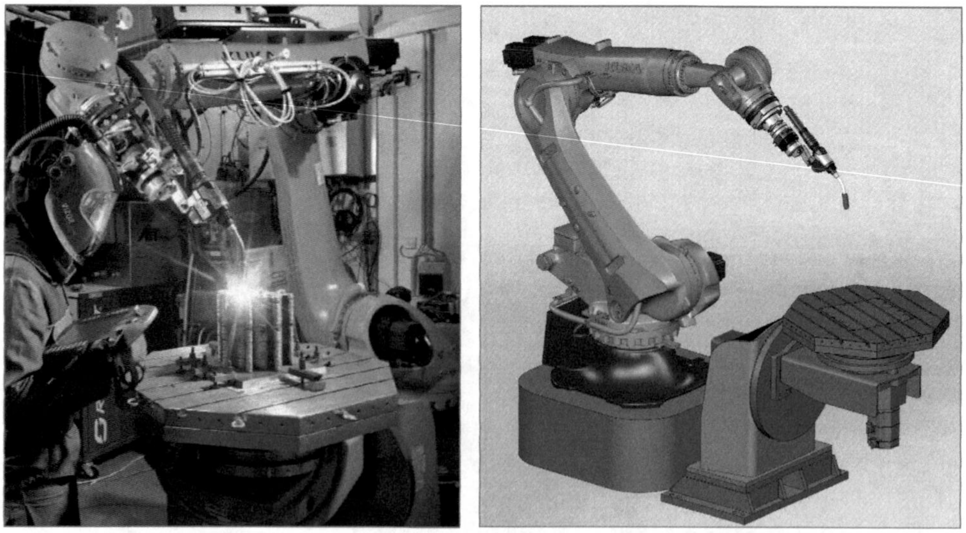

Figure 2. KUKA Robot setup with tilt-turn table and Fronius welding power source (Left); Offline robot programming and simulation in Robotmaster software (Right).

2.2 Materials

Argon and Argon-Helium mixtures are used as shielding gases for the study. Argon of 99.999% purity grade and Varigon 30 (Ar70+He30) are used. The feedstock material used is 99% copper alloy (CuSn1) in 1.2 mm wire form and the substrate material is a high conductivity phosphorous copper (Cu-HCP) plate of 10mm thickness. CuSn1 has good weldability compared to pure copper using gas metal arc welding. The compositions and properties of CuSn1 are described in the following tables.

Table 2. CuSn1 alloy composition in mass percentage.

Alloy	Copper	Tin	Silicon	Manganese	Ni+Co, Pb, Fe, P, Others
CuSn1 S Cu 1898A (CuSn1MnSi)	99 %	0.5-1.0	0.10-0.40	0.10-0.40	< 0.10, < 0.01, < 0.03, < 0.015, < 0.20

Table 3. Material Properties of CuSn1 alloy.

Alloy	Melting Point (°C)	Thermal Cond. (W/m.K)	Electrical Cond. (%IACS)	Tensile Strength (MPa)	Elongation (%)
CuSn1	1050	120-140	25-34	210 – 245	30

2.3 Process parameters

Process parameters are analyzed from literature research to be in the following ranges for GMAW: gas flow rate of 15-20 L/min, travel speed of 6-8mm/s (approx. 2mm/s for TIG) and wire feed speed of 4-6 m/min. Welding standards like AWS welding handbook and the Copper Development Association recommend pre-heating temperatures of 300-500 °C for welding of copper substrates with thickness of 10mm or more. Helium gas is recommended for shielding due to higher heat in the arc compared to Argon gas. This is due to the higher ionizing temperature and higher thermal conductivity in ionized state of Helium gas. A mixture of both gases provides the arc stability of Argon and higher heat from Helium [6, 7, 8]. The information combined with welding technology and WA-DED concepts are utilized to setup process parameters for the copper WA-DED process.

3. Investigation of Copper WA-DED

The methodology followed for the study described in brief:

1. Initial trials to achieve stable deposition parameters.
2. Production of coupons.
3. Metallography test – porosity analysis and microstructure evaluation.
4. Vickers micro-hardness testing.
5. Process optimization and production of demonstrator part.

The geometries produced for this study are trial walls, billets and a hexagonal thin-walled test part. The initial trials are with single beads of weld for ascertaining a stable parameter and acceptable layer geometry. The billets are produced for evaluating the manufacturability of solid copper parts and as test specimens for future electrical conductivity testing. The geometries used in the study are shown in **Figure 3**.

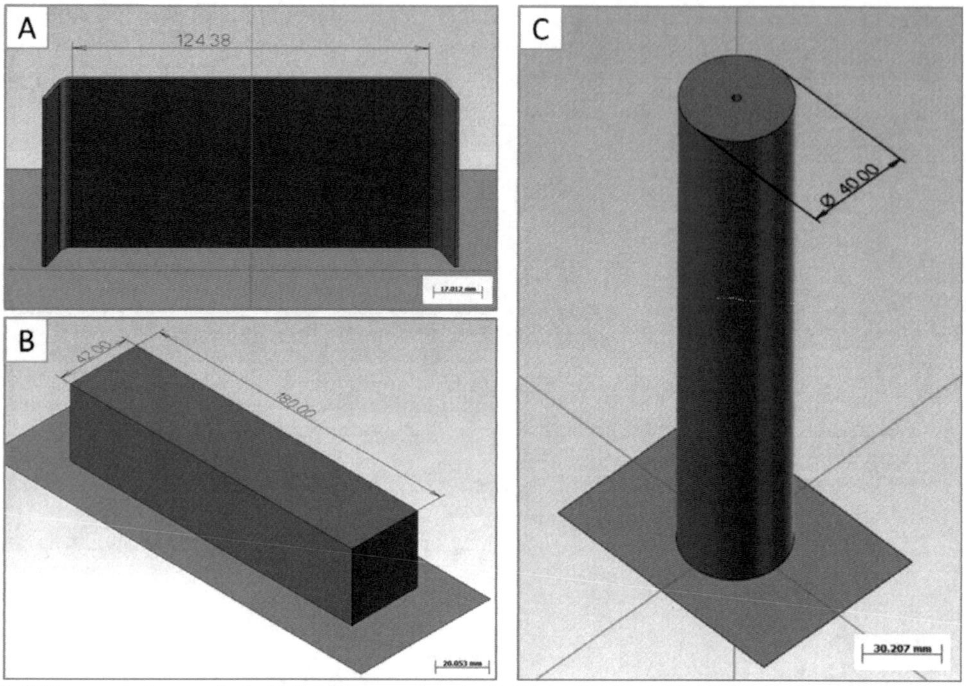

Figure 3. CAD geometries used - A: Wall with curved ends, B: Horizontal cuboidal billet, C: Vertical cylindrical billet.

3.1 Walls

The walls are designed with curved start and end points to reduce the layer height difference commonly observed with straight tool paths. The CAM programming is a one-way toolpath with and without robotic weaving motion. The initial trials show that a preheating of 75°C is insufficient for welding the copper alloy on a 10mm Cu-HCP substrate even with high heat input process parameters.

The trials with spiral weaving motion combined with pre-heating of 250°C produced a wider weld bead and improved joining with the substrate. The spiral weaving motion increases the local heat input during welding and improves the stability of the melt pool. However, the fusion with the substrate is not satisfactory at the starting of the weld track up to 40 mm length. A pre-heating temperature of 300°C improves the integration with the substrate and a stable melt pool is maintained during layer deposition.

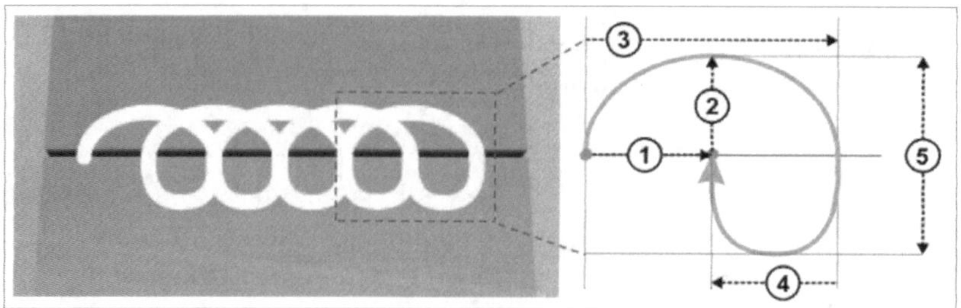

Figure 4. Spiral weaving motion using Kuka Arctech Advanced – 1: Weave length (half of (3)), 2: half amplitude of weaving (half of (5)), 4: reverse travel of torch in spiral weaving (equals (1)). Source: Kuka Arctech Advanced Manual

The final walls are produced after parameter optimization (**Figure 5**). Two walls are produced using CMT and pulse WA-DED processes, respectively. Metallographic samples are extracted from the walls to evaluate the porosity levels and the grain structure in the deposited metal. The results of these tests will be discussed in section 4.

Figure 5. Walls produced with CMT and pulse WA-DED processes.

3.2 Billets

The billets are produced both horizontally and vertically. The toolpath is a zig-zag strategy with a fixed layer stepover and without weaving motion. The direction of tool travel is reversed for each layer to maintain uniform material distribution and hence uniform layer height.

The base layer toolpath for the vertical billet is an additional path around the actual billet geometry (**Figure 6**A Top). It is designed to heat the local substrate volume and aid in the melt pool creation. A spiral toolpath directed towards the inside of the cylinder is programmed to retain the heat in the weld region. This also has a spiral weaving motion of the torch programmed. The toolpaths are shown in **Figure 6**A.

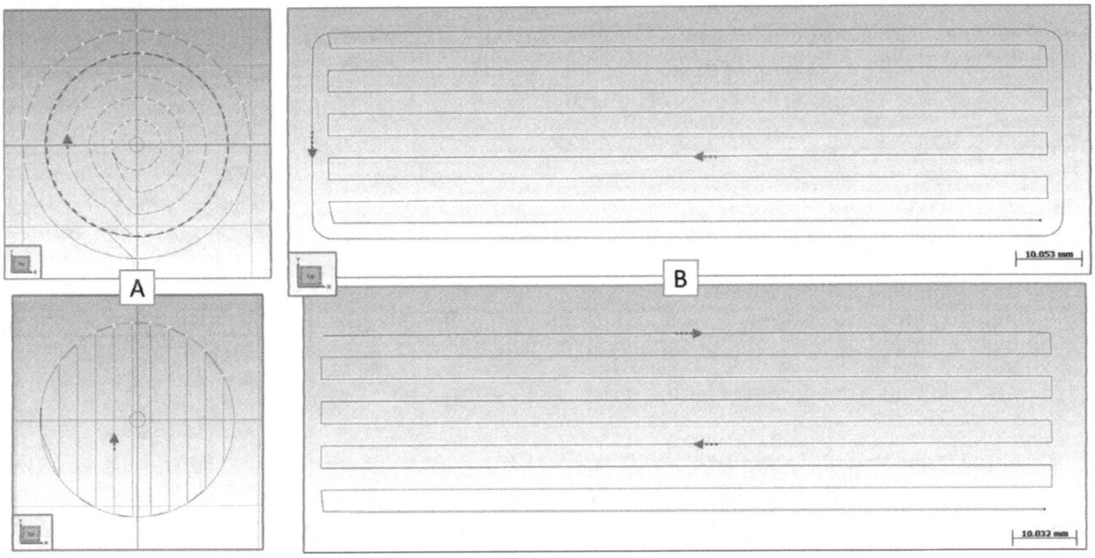

Figure 6. CAM toolpath strategy for vertical cylindrical billet(A) and horizontal cuboidal billet - Base layer (A, B: Top) and rest of the layers (A, B: Bottom, only a layer shown).

The base layer of the horizontal billet is started with a ring weld offset outwards to the dimensions of the billet (**Figure 6**B Top). The goal is to stop the lack of fusion occurring at the start of weld track to be included in the billet part. The toolpath is planned parallel to the longer side of the billet to not overheat the local region and distribute the heat along the length of the part. The toolpaths are shown in **Figure 6**B.

Figure 7. Solid billets (A: Vertical, B: Horizontal) produced using CuSn1 alloy.

The billets are produced with a pre-heating temperature of 280°C and a welding speed of 6 mm/s. The part is observed to get red hot during WA-DED processing and cool down extremely fast when the process is interrupted. A high heat retention due to continuous welding leads to a bigger melt pool and, at times, can cause melt flow at the edges of the geometry. Conversely, a long wait time between passes leads to cold weld start and improper fusion with the previous layer. Therefore, the inter-pass cooling time is not fixed, and it is based on in-situ process monitoring of the size and stability of the melt pool. This helps in avoiding any catastrophic melt flow on the edges of the billet. The billets produced by WA-DED are shown in **Figure 7**.

4. Results and Discussion

The results of the metallography and micro-hardness testing are discussed in this section.

4.1 Porosity Analysis and Microstructure

The porosity analysis of the samples is conducted, and the resulting images are shown in **Figure 8**. Porosity levels are determined to be less than 0.01% of the observed surface area and the rare pores observed have an average diameter below 20 µm.

It is also observed that the porosity levels for CMT and Pulse modes are equivalent which can be observed in **Figure 8** between 1,2 and 3,4. The results aid in concluding that the processing parameters are optimal for the WA-DED of CuSn1 alloy to achieve a relative density of 99.99%.

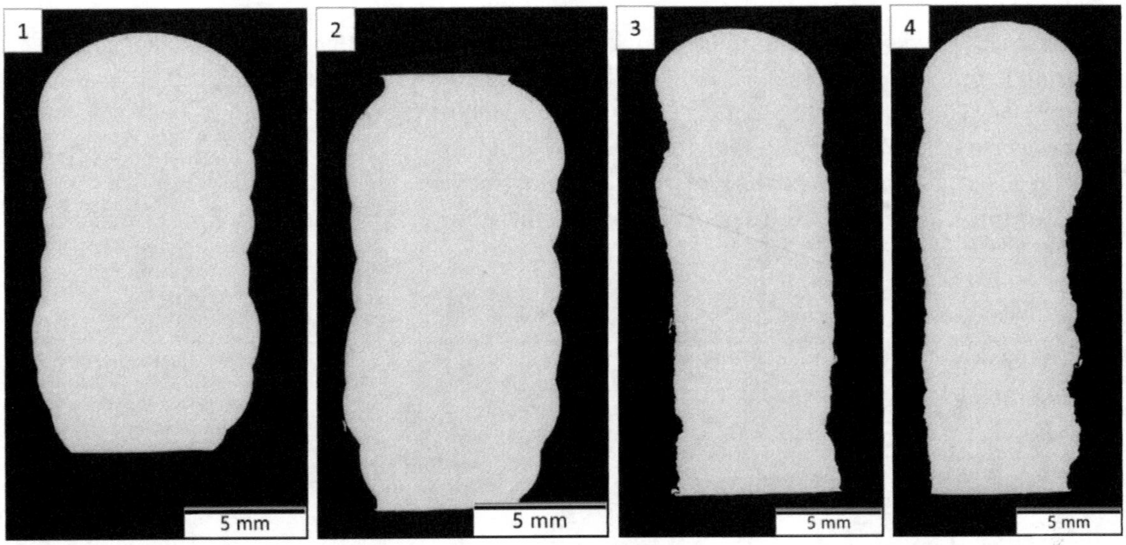

Figure 8. Images from metallographic samples from wall for porosity analysis - 1,2: CMT process; 3,4: Pulse process.

The specimens are prepared for microstructure observation by etching with a solution of HCl, Fe(III)Cl and distilled water. The resulting metallographs are shown in **Figure *9***. The lower end of the samples is observed to have smaller grains compared to the top region. This is due to the distance from the substrate and the reduction in heat dissipation with repetitive deposition and heat accumulation in the part. The samples produced by CMT process (1) have smaller average grain size compared to those produced by pulse process (2). This is likely because of the lower heat input in the CMT process [9]. In the pulse process samples, distinct large columnar grains are observed owing to the epitaxial growth mechanism in the layer-upon-layer deposition process. No anomalies are observed at the grain boundaries or elsewhere.

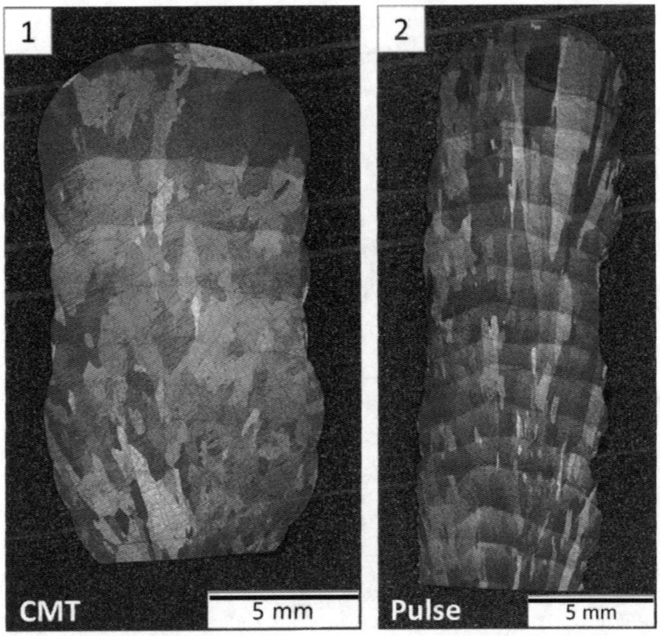

Figure 9. Etched metallographs of the samples oriented along build direction.

IOP Conf. Series: Materials Science and Engineering **1315** (2024) 012004 doi:10.1088/1757-899X/1315/1/012004

4.2 Hardness Tests

The micro hardness on the samples from produced walls is evaluated using Vickers hardness tester according to EN ISO 6507-1. The $HV_{0.1}$ is conducted with a load value of 0.98 N and load time of 10 s. The imprints are made at every 1mm along the build direction (Z-axis). The results are summarised in **Table 4** and **Figure 10.**

The average micro-hardness for the CuSn1 alloy samples is 64 $HV_{0.1}$. No trends are observed along the build direction of the samples. Additionally, no significant difference between CMT and pulse process modes is observed.

Table 4. Micro-hardness results for CuSn1 alloy WA-DED samples.

Process	CMT	Pulse	Overall
$HV_{0.1}$ Average, Std. Dev.	64.1 1.35	64.4 2.10	64.3 1.81

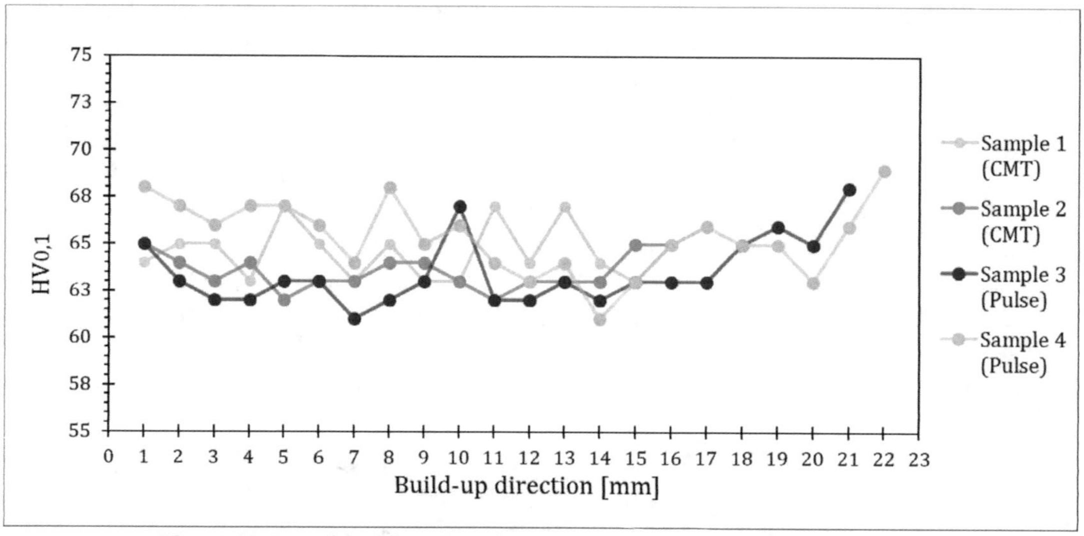

Figure 10. Data plot of the $HV_{0.1}$ values at every imprint point for four samples.

5. Demonstration Part for e-Mobility Application

Busbars are critical electrical components used for transmitting electrical power from a supply to sub-circuits. They are designed to minimize thermal losses and improve the efficiency of electrical systems including e-mobility systems. A busbar geometry is used as a demonstration of the WA-DED process using copper.

The geometry and CAM planning for the busbar is shown in **Figure 11**. It is produced with the base of the geometry as a part of substrate and the two separated sections (1, 2) built-up using WA-DED. The CAM programming consists of zig-zag toolpath along the longer length to distribute the heat input uniformly and avoid melt flow at the edges of the small deposition area. For the first section (1), the welds are started outside the part area to avoid lack of fusion with substrate based on previous learnings. The second section of the geometry (2) is made using 90° tilt of the robotic table in the same way. A pre-heating of 240°C is used and an inter-pass cooling time is regulated, as necessary. The busbar is produced using Varigon30 shielding gas and the parameters developed for WA-DED. The as-built part is shown in **Figure 12**.

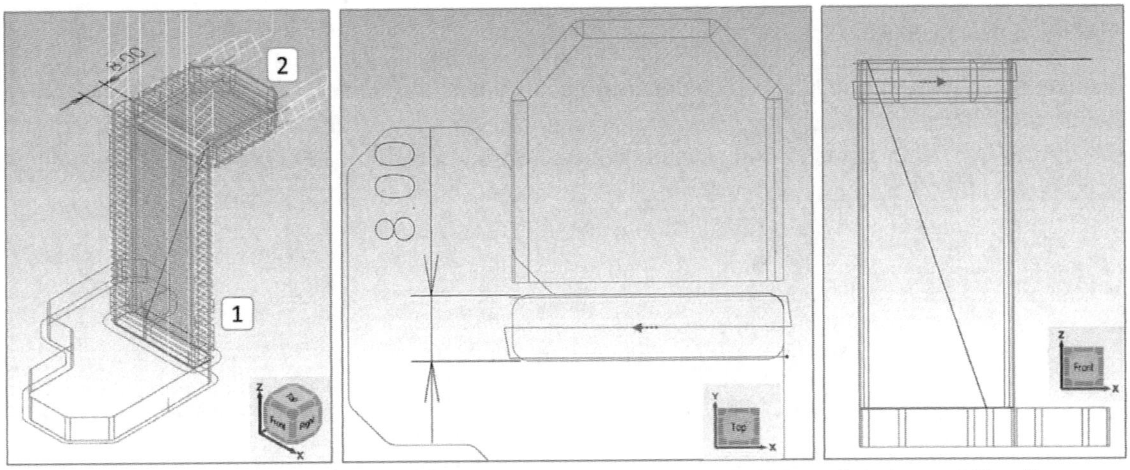

Figure 11. CAM Toolpath for the Busbar part- Left: Overview, Middle: first wall, Right: top 90° wall.

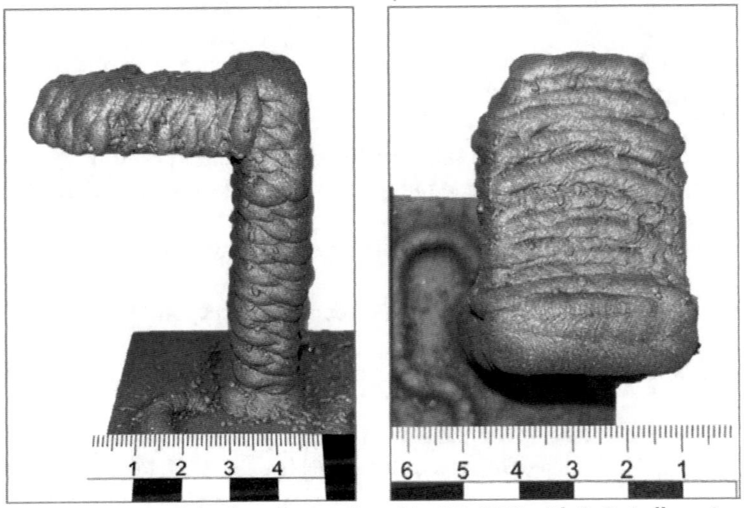

Figure 12. As-built busbar part produced by WA-DED with CuSn1 alloy wire.

6. Conclusions and Future Outlook

The trials in the study show the feasibility of Copper WA-DED for thin wall and solid parts. Lack of fusion caused by fast thermal dissipation due to the high thermal conductivity of copper is a major challenge during WA-DED. It is found that a pre-heating temperature of 300°C is suitable for processing the CuSn1 alloy with Argon gas shielding and 250°C for processing with Argon70Helium30 gas shielding. Geometrical consistency of the deposited layers is observed to be better in CMT process compared to Pulse process. The WA-DED process parameters developed for CuSn1 achieve a relative density of 99.99% and an average $HV_{0.1}$ of 64 in the part.

The research with copper WA-DED are ongoing and mechanical and electrical resistivity testing will be carried out using the produced parts. The demonstration of WA-DED processing of copper busbar shows promise for for e-mobility applications. The future investigations will be focusing on evaluating the process feasibility and properties in the part for pure Copper WA-DED.

7. Acknowledgements

This research was funded by the federal state of Upper Austria in the project "GEPROBA - Gesteigerte Prozess- und Bauteilperformance durch neue Fertigungsmöglichkeiten für eMobilitäts-Anwendungen" (grant no. 901696) administered by FFG within the framework "Future Mobility". The authors thank the project partners for the collaborative work.

References

[1] Copper Development Association (CDA) www.copper.org, Accessed in April 2024

[2] B Baufeld Wire electron beam additive manufacturing of copper 2023 J. Phys.: Conf. Ser. 2443 012001

[3] S. W. Williams, A. Rashid G. Pardal & L. Quintino WIRE AND ARC ADDITIVE MANUFACTURE OF HIGHLY CONDUCTING PURE COPPER, RAPDASA 2019 PROCEEDINGS;01:1–4.

[4] Baby J, Amirthalingam M. Microstructural development during wire arc additive manufacturing of copper-based components. Weld World 2020; 64:395–405.

[5] Poonam S. Deshmukh, Krishna Tomar, G. Dan Sathiaraj, I.A. Palani, Optimum strength and ductility of pure copper fabricated by Wire Arc Additive Manufacturing, Manufacturing Letters, Volume 33, 2022

[6] Joining of Copper and Copper Alloys, 1994, Copper Development Association (CDA)

[7] Copper for Busbars, 2014, Copper Development Association (CDA) and European Copper Institute

[8] Welding Handbook Volume 5 (Materials and Applications Part 2), American Welding Society (AWS)

[9] Jicai Feng, Hongtao Zhang, Peng He, The CMT short-circuiting metal transfer process and its use in thin aluminium sheets welding, Materials & Design, Volume 30, Issue 5, 2009, Pages 1850-1852

[10] Das, Abhishek & Barai, Anup & Masters, Iain & Williams, (2019). Comparison of Tab-To-Busbar Ultrasonic Joints for Electric Vehicle Li-Ion Battery Applications. World Electric Vehicle Journal. 10. 55.

[11] Sebastian Engler, Reiner Ramsayer, Reinhart Poprawe, Process Studies on Laser Welding of Copper with Brilliant Green and Infrared Lasers, Physics Procedia, Volume 12, Part B, 2011, Pages 339-346

IOP Conf. Series: Materials Science and Engineering **1315** (2024) 012005 doi:10.1088/1757-899X/1315/1/012005

Re-manufacturing of Pre-deformed Automotive Sheet Metal Stamping Scrap Without Melting

Johannes A. Österreicher*, Florian Grabner, Maria Kühlein, Angelika Cerny, and Carina M. Schlögl

LKR Light Metals Technologies, AIT Austrian Institute of Technology, Ranshofen 5282, Austria

*E-mail: johannes.oesterreicher@ait.ac.at

Abstract. The recycling of aluminium sheet metal typically involves shredding and re-melting, processes which are both energy-intensive and challenged by the presence of metallic contaminants. To mitigate these issues, this study explores an alternative approach: the re-manufacturing of deep-drawn parts directly from scrap aluminium sheets, thus bypassing the melting step. We used 2 mm thick Al-Mg sheet metal in H111 temper to produce simulated stamping scrap (miniaturized bonnets). Blanks were cut from these bonnets and secondary parts in a cross-cup geometry were deep-drawn using two different forming processes: warm-forming at 200 °C and O-temper forming at room temperature. The resulting maximum draw depths were 20–22 mm for warm-forming and 25–30 mm for O-temper forming. While warm-forming resulted in lower draw depths, O-temper forming led to coarse recrystallized grain formation in some regions. Our approach could significantly reduce the energy consumption associated with aluminium recycling.

1. Introduction

Sheet metal scrap is typically shredded and re-melted [1]. The homogeneity of the shredded metal depends on the sorting operations conducted before or after shredding; it can be relatively pure or contaminated [2]. Non-metallic contaminants can be removed through combustion or melt-cleaning (raffination) procedures [3]. However, metallic tramp elements, such as iron in aluminium melts, are challenging to eliminate [4], [5]. Furthermore, even with optimal sorting and cleaning operations, the re-melting process has an unavoidable energy requirement due to the enthalpy of fusion.

Aluminium recycling is widely recognized to require only 5–10% of the energy needed for primary production via electrolysis. According to Fragner [6], the enthalpy of fusion constitutes 47% of the energy requirements in the traditional aluminium recycling process, while scrap preparation, raffination, and dross reworking account for another 47%. The remaining 6% is attributed to transportation. Additional energy is required for rolling new aluminium sheets. Consequently, eliminating both the melting and rolling steps could significantly lower the energy consumption associated with recycling aluminium alloys. This paper will examine the production

(a)

(b)

Figure 1. Schematic of blanking scrap, stamping scrap, and part in stamping of an automotive body side from sheet metal coil (schematic based on [2]). (b) EoL bonnet made from Al alloy sheet (disassembled from Audi A6 4B, year 2000, 240,000 km).

of deep-drawn parts from scrap aluminium sheets using a re-manufacturing process that avoids the intermediate melting step.

Regarding aluminium sheet material used in the automotive industry, scrap can be broadly categorized into three types: first, undeformed blanking scrap (cutoff); second, stamping scrap (deformed cutoff from parts and defective deep-drawn components); and third, end-of-life (EoL) parts (e.g., auto bonnets) [7], see Figure 1. Blanking scrap retains the characteristics of virgin sheet metal, making it potentially straightforward to use for producing smaller parts. Stamping scrap, on the other hand, differs from the raw material due to its pre-existing deformation, which affects its strength and formability. EoL scrap presents the greatest challenge, as it requires separation from other materials and possibly removal of paint [8]. Additionally, this type of scrap often lacks detailed information about the original materials and their life-cycle, including load cycles and corrosion damage.

Literature on sheet metal re-manufacturing (or re-forming) is relatively scarce. Abdullah [9] addressed the growing environmental burden of automotive industry waste by exploring a small business model that remanufactures metal sheets from end-of-life vehicles into wall junction boxes for construction. The study evaluates the technical, economic, and environmental feasibility of this remanufacturing process using commercially available technologies, showing high sustainability and substantial profitability across different production scenarios.

In a study by Copani et al. [10], laser de-soldering for disassembly and the production of non-decorative interior parts from an EoL car roof achieved a 40% reduction in CO_2 footprint. Challenges were identified in both technology (improving part quality and process scalability) and business models (such as establishing reverse chains, ensuring customer acceptance, and pricing).

Horton and Allwood [7] investigated the feasibility of using blanking scrap, stamping scrap, and EoL car parts to produce new components, potentially lowering costs and reducing environmental impact. They found that while some yield losses are unavoidable with current technologies, most organizations have the potential to enhance material utilization.

In this paper, we investigate the production of secondary deep-drawn parts from stamping scrap. For this purpose, we first manufacture miniaturized bonnets as simulated stamping scrap (rejected products). Then, secondary parts are produced in a cross-cup geometry. Maximum draw depth, optical appearance and grain structure are assessed.

2. Materials and methods

We used 2 mm thick AA5182 and AA5083 sheet metal in H111 temper and stamped miniaturized bonnets (Figure 2) using a 250 t servo-hydraulic drawing press. The bonnet (engine hood) geometry was used merely for demonstration purposes, as 5xxx series alloys (Al-Mg) are typically only used for automotive inner parts due to the occurrence of stretcher-strain marks [11].

From these parts, blanks were water jet cut for the use with a cross-die tool (Fig. 3). These secondary blanks were either pre-heated in a furnace at 200 °C and deep-drawn in the heated tool (also 200 °C); or O-temper treated (400 °C for 2 h, cooled at 50 K/h) and deep-drawn into cross-cups at room temperature (RT). The schematic process chains are shown in Figure 4.

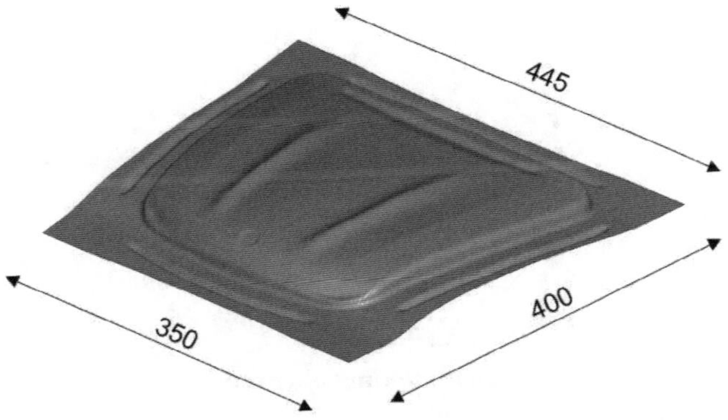

Figure 2. Isometric view of miniaturized bonnet stampings. Dimensions in mm.

Figure 3. Secondary blanks were water jet cut from various positions of the stamped parts.

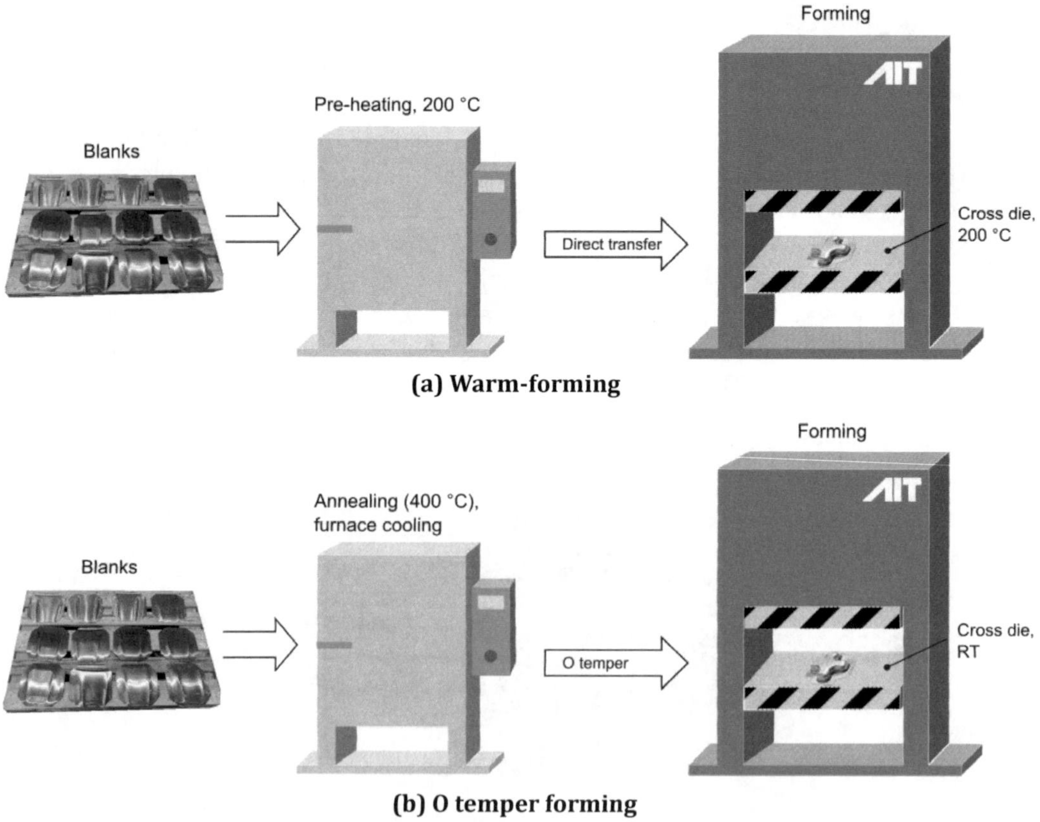

Figure 4. Remanufacturing by (a) warm-forming and (b) O temper forming.

By re-manufacturing of deep-drawn cross-cups, the maximum draw depth was assessed. The deep-drawn parts were visually inspected and metallographic cross-sections were taken to evaluate possible coarse grain formation by optical light microscopy after etching with Barker's reagent.

IOP Conf. Series: Materials Science and Engineering **1315** (2024) 012005 doi:10.1088/1757-899X/1315/1/012005

3. Results and discussion

The deep-drawing of re-manufactured parts resulted in maximum draw depths of 20–22 mm for the warm-forming process and 25–30 mm for the O temper forming process, as illustrated by the example part in Figure 5. No significant difference was observed between the two alloys tested, AA5083 and AA5182. The lower draw depth achieved in the warm-forming process, compared to the O temper forming, is likely due to the material softening at 200 °C. This softening leads to earlier necking and, consequently, fracture.

Figure 5. Example of lubricated secondary blank in O temper (left) and cross-cup deep-drawn from it (right). Bonnet character lines marked with arrows.

The surface quality of the re-manufactured parts was generally acceptable; however, the bonnet character lines (i.e., edges that serve both aesthetic and stiffening functions) remained visible on the parts. Therefore, re-manufactured parts may be more suitable for non-exterior applications in automaking [10] or for other non-aesthetic components.

Metallographic characterization of the original bonnet character line region in the re-manufactured part revealed coarse recrystallized grain for the parts formed via the O temper forming process (Figure 6), while no recrystallization was observed for warm-forming at 200 °C. Depending on the application, such coarse grain formation may be unacceptable [12]. In these cases, either warm-forming could be employed or the character lines could be recessed during the secondary blanking process.

Figure 6. Metallographic section of original bonnet character line region in a re-manufactured part (cross-cup, O temper forming).

4. Conclusions

We investigated re-manufacturing of deep-drawn parts from stamping scrap (i.e., pre-deformed material from another deep-drawing operation). The following conclusions can be drawn:

- Both O temper forming and warm-forming are suitable processes for production of re-manufactured parts from 5xxx series alloys.
- O temper forming achieved higher draw depths and does not require heated tools, but coarse grain formation can occur in highly deformed regions of the primary part (stamping scrap) during annealing.
- Regardless of re-manufacturing process used, highly deformed features of the primary parts (e.g., body character lines) are still visible on the re-manufactured parts, which therefore are deemed more suitable for non-visible applications.

5. Acknowledgments

Funding: Financial support by the Austrian Research Promotion Agency (FFG) is gratefully acknowledged (ProMetHeus, FFG no. 904919).

References

[1] M. E. Schlesinger, *Aluminum Recycling*. CRC Press, 2014.
[2] P. Krall, I. Weißensteiner, and S. Pogatscher, "Recycling aluminum alloys for the automotive industry: Breaking the source-sink paradigm," *Resources, Conservation and Recycling*, vol. 202, p. 107370, Mar. 2024, doi: 10.1016/j.resconrec.2023.107370.
[3] R. Gallo, "Cleaner Aluminum Melts in Foundries: A Critical Review and Update," in *AFS Transactions*, 2008, pp. 195–220.
[4] Q. Li *et al.*, "Accelerating the dissolution and transformation of detrimental Fe-rich intermetallics in recycled Al-Mg-Si alloy by homogenization and hot deformation," *Materials Characterization*, vol. 213, p. 114024, Jul. 2024, doi: 10.1016/j.matchar.2024.114024.

[5] J. A. Österreicher, A. R. Arnoldt, S. Gneiger, and G. Kunschert, "Tolerance of Al–Mg–Si Wrought Alloys for High Fe Contents: The Role of Effective Si," *Metallurgical and Materials Transactions A*, vol. 54, pp. 4472–4480, 2023, doi: 10.1007/s11661-023-07180-z.
[6] W. Fragner, "Herausforderungen und Chancen der Aluminium-Prozesstechnologie im europäischen Umfeld," in *Proceedings of 8. Ranshofener Leichtmetalltage*, Geinberg, 2014.
[7] P. M. Horton and J. M. Allwood, "Yield improvement opportunities for manufacturing automotive sheet metal components," *Journal of Materials Processing Technology*, vol. 249, pp. 78–88, Nov. 2017, doi: 10.1016/j.jmatprotec.2017.05.037.
[8] H. Geist and F. Balle, "A circularity engineering focused empirical status quo analysis of automotive remanufacturing processes," *Resources, Conservation and Recycling*, vol. 201, p. 107328, Feb. 2024, doi: 10.1016/j.resconrec.2023.107328.
[9] Z. T. Abdullah, "Remanufacturing waste steel sheet from end-of-life vehicles into electrical installation wall junction boxes: Quantitative sustainability assessment," *Results in Engineering*, vol. 21, p. 101767, Mar. 2024, doi: 10.1016/j.rineng.2024.101767.
[10] G. Copani, P. Shafinejad, T. Hipke, R. Haase, and T. Paizs, "New metals remanufacturing business models in automotive industry," *Procedia CIRP*, vol. 112, pp. 436–441, Jan. 2022, doi: 10.1016/j.procir.2022.09.033.
[11] F. Grabner, J. A. Österreicher, *et al.*, "Cryogenic Forming of Al-Mg Alloy Sheet for Car Outer Body Applications," *Advanced Engineering Materials*, vol. 21, no. 8, p. 1900089, 2019, doi: https://doi.org/10.1002/adem.201900089.
[12] P. Goik, A. Schiffl, H. W. Höppel, and M. Göken, "The Role of Through-Thickness Variation of Texture and Grain Size on Bending Ductility of Al–Mg–Si Profiles," in *Light Metals 2023*, 2023, pp. 568–577. doi: 10.1007/978-3-031-22532-1_77.

IOP Conf. Series: Materials Science and Engineering **1315** (2024) 012006 doi:10.1088/1757-899X/1315/1/012006

Enhancement of the electrical conductivity of APPD manufactured copper conductors integrated into aluminium WAAM structures

R Gradinger[1,2*] and D Hohm[1]

[1] LKR Leichtmetallkompetenzzentrum Ranshofen GmbH, Ranshofen, Austria
[2] Chair of Materials Engineering of Additive Manufacturing, TUM School of Engineering and Design, Technical University of Munich, Garching, Germany

*E-mail: rudolf.gradinger@ait.ac.at

Abstract. Copper conductors - manufactured by atmospheric plasma powder deposition, insulated by alumina layers and integrated into aluminium structures - were analysed with regard to their electrical conductivity. Variations of a novel electrical post-processing were studied in comparison to standard heat treatment. Both of them showed significant improvement of the specific electrical resistivity.

1. Motivation

The variety of additive manufacturing (AM) technologies has already proven its major factors of benefits, first of all the great freedom of design of complex shaped parts.
Another important opportunity is the application of several different materials – even material classes - in one part in order to generate novel functionalities. Multi-material design enabled by Direct Energy Deposition (DED) technologies was in the heart of the EU funded project MULTI-FUN - focussed on a variety of integrated functionalities demonstrated by seven real size use case related parts [1].

One of these demonstrators - the motorcycle handlebar - aimed at the integration of all basic electric and hydraulic signal transfer features (for the control of clutch, front brake, starter, indicator, lights as well as horn) into the mechanical structure without using visible cables (Figure 1). While the design of the hydraulic channels resembled cooling channels well known from additively manufactured moulds, the full integration of electrical power transfer needed a novel approach. This approach consisted of manufacturing the mechanical structure of the handlebar in a stepwise procedure by an appropriate AM method, interrupted by the process of depositing layers of ceramic (functioning as electrical insulator between all neighbouring metal volumes) as well as pure copper (as the standard electrical conductor material), resulting in five distinct copper tracks, see Figure 2.

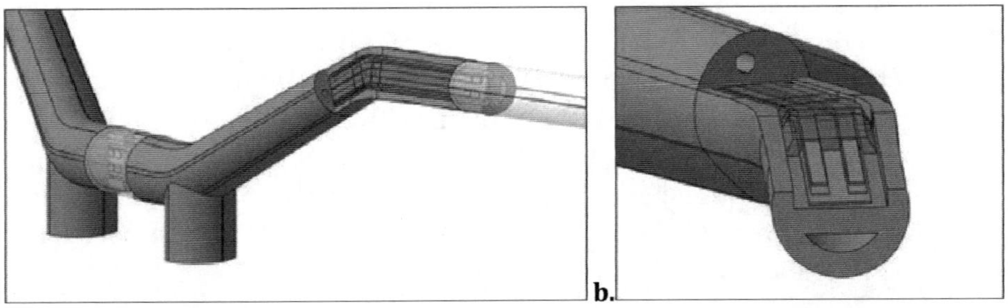

a. b.

Figure 1. Generic design of the multi-material motorcycle handlebar; a.) overall shape; b.) details of multi-functional design with copper lines (orange colour), alumina insulators (light blue) and the hydraulic channel shown as a hole in the aluminium structure (grey), drawings courtesy of EDAG [1]

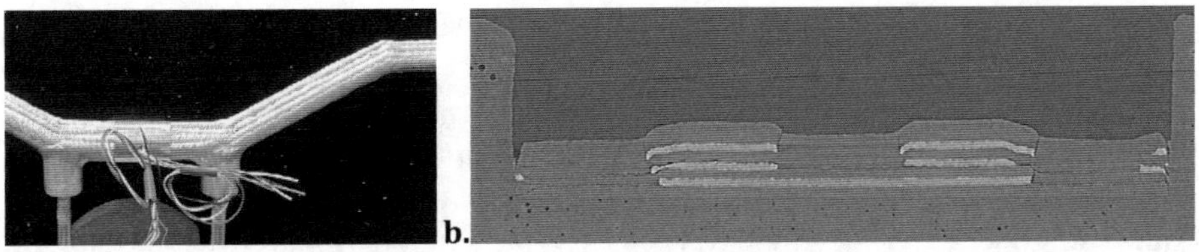

a. b.

Figure 2. a.) WAAM based assembly of the motorcycle handlebar; b.) Cross section of the insulator and five copper conductor stacks, both manufactured by APPD onto an aluminium substrate before WAAM integration, corresponding to Figure 1.b. (copper tracks in light grey; alumina deposited all around the copper tracks, image processed by DLR) [2]

The two main DED methods chosen were Wire-Arc Additive Manufacturing (WAAM) as well as Atmospheric Plasma Powder Deposition (APPD) because of their ability to create 3D structures out of aluminium wire (for the mechanical structure), copper powder (as the typical electrical conductor) as well as alumina powder (Al_2O_3 acting as the insulator in between aluminium and copper elements). Before the demonstrator manufacturing was started, principal investigations about the prove of concept of insulation of copper tracks and their resulting conductivity and the mechanical robustness of the multi-material design were performed [3].

Early investigations showed that the electrical resistivity of copper tracks manufactured by APPD was by a factor of appr. four times higher as the standard properties of pure copper, widely stated at a level of 0,017 $\Omega \cdot mm^2/m$.

A hypothesis about the main mechanisms causing the poor electrical conductivity of as-deposited copper tracks was put forward which included the influences from oxygen solved in copper of high purity, the nano-sized grain structure within each powder particle as well as the interfaces of the deposited micro-sized powder particles to each other.

A concept for a novel special process developed based on the application of an electrical current to the copper tracks, resulting in a quick temperature increase according to Joule's law. It states that the generated heat "H" in Joule (J) is directly proportional to the product of electrical resistance "R" in Ohm (Ω) times the time "t" of the electrical current flowing in seconds (s) times the square of the current "I" in Ampere (A). The formula reads as $H \sim R \times t \times I^2$. This new post-treatment approach resembled to a certain degree the principles of Spark Plasma Sintering (SPS) and its variations known as Field Activated Sintering Technique (FAST) or Plasma Activated

Sintering (PAS). SPS usually applies low-voltage and pulsed direct current combined with low mechanical pressure aiming at a fast consolidation of preconsolidated powder. A common theory involves electrical discharge at small contact areas of adjacent powder particles, generating a plasma spark at high temperatures within a fraction of a second [5].

The changes to the nano-, micro- and macro-structures of the copper tracks by these sparks and the related Joule's heat were considered to sustainably improve the electrical properties. The effects of such procedures in comparison as well as in combination with standard heat treatment (HT), respectively, at certain stages of the process chain of WAAM were reported extensively for the first time in [4]. The main findings are reported in the following chapters.

2. Sample design

A simplified sample design was chosen for principial investigations about the influence of electrically induced post-processing. It was exhibiting the basic features of the future demonstrator part but was reduced in size, in geometrical complexity as well as in number of insulated conductor stacks.

The aluminium substrates were extracted from an extruded profile of standard EN AW-6060 alloy (AlMgSi0,5). The cross-section of the copper layers was defined to be 5 mm x 150 µm. The length of the copper layer was 130 mm, deposited upon a centred, 10 mm wide section of alumina, which is spreading along the full substrate length. Most (but not all of the samples incl. the one shown in Fig.6) were again treated by APPD to produce an insulting alumina layer above the conductor strip in order to prevent short circuiting after subsequent integration into aluminium. For later processing (incl. contacting), both ends of the copper stack were kept uncoated for about 5 mm length to maintain the conductivity (see Fig.5.a). The drawing of this generic sample design is shown in Figure 3.

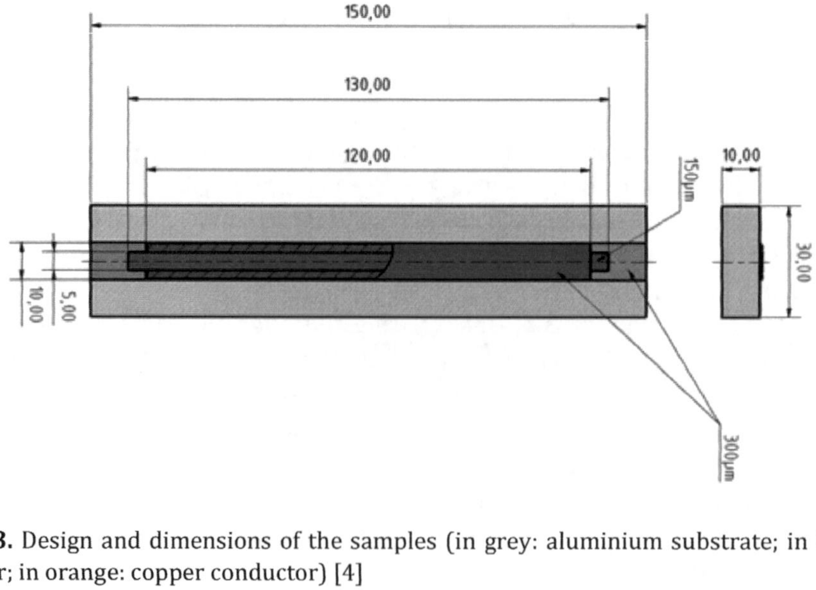

Figure 3. Design and dimensions of the samples (in grey: aluminium substrate; in blue: alumina insulator; in orange: copper conductor) [4]

3. Manufacturing of the samples

The extrusions with a thickness of 10 mm were at first sand blasted on the surface dedicated for additive manufacturing with the aim of enhancing the adhesion of the first ceramic layer. A handheld device was applied with parameters established by experience.

Thermal spraying was performed by an APPD equipment of type InoCoat IC3, operated by the manufacturer INOCON [6]. A masking was applied to cover the areas where no material should be deposited. Commercially available copper powder with a d_{50}=9,89 µm was used for the conducting layers. For the ceramic insulation layers, alumina powder (Al_2O_3) with a d_{50}= 9,3 µm was applied. Figure 4 is showing a representative cross section of a copper tracks insulated by Al_2O_3.

The integration of the insulated conductor stack was performed in LKR's additive manufacturing laboratory, applying standard FRONIUS CMT welding equipment as well as conventional welding wire MA-5183 (AlMg4,5Mn0,7). Stepwise, the following DED processes were applied:

1.) Thermal spraying of 10 mm wide, appr. 200 µm high layered stack of alumina along the middle of the substrate bar, along its full length of 150mm,
2.) Thermal spraying of 5 mm wide, appr. 150 µm high layered stack of copper along the middle of the alumina strip, limited to a length of 130 mm,
3.) Covering the layered stack of copper by alumina, leaving both end of the copper strip (of appro. 5 mm) uncoated (in order to act as the contact areas), see Fig. 5. a.
4.) Integration of the insulated conductor stack by wire arc additive manufacturing in the middle of the samples, stopping left and right at a distance of about 5 mm length to the uncoated (in order to avoid any negative to the contact areas), see Fig. 5. b.

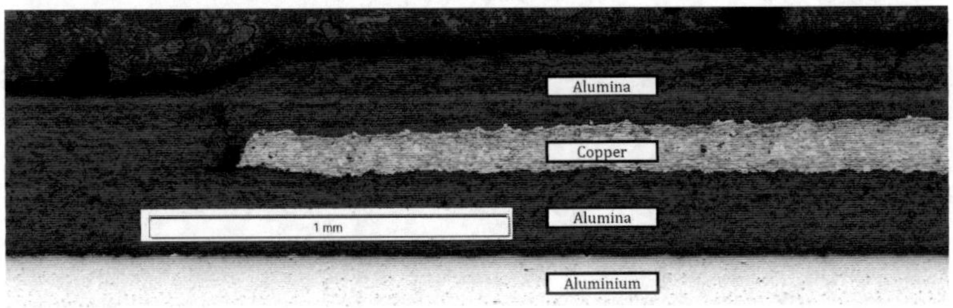

Figure 4. Optical microscopy offering an insight into the grain structure of the alumina and copper stacks .

4. Post-processing

Both the sample substrate as well as the designated wire electrode for the motorcycle handlebar demonstrator, namely MA-6063 (AlMg0,7SiTiB), are reacting to heat treatment (HT) [7]. Therefore, the standard T6 heat treatment of an EN AW-6060 alloy was applied to a part of the samples, applying a temperature of 540°C for 1 h plus subsequently 180°C for 8 h using a conventional heat treatment furnace. The influence of this heat treatment to the electrical resistance properties of the integrated copper tracks was of main interest in comparison to the effects of a novel electrical treatment. This post-processing concept was studied in two variants, on the one hand a short-time procedure with a high current density of up to 1000 A/mm^2

a.
b.

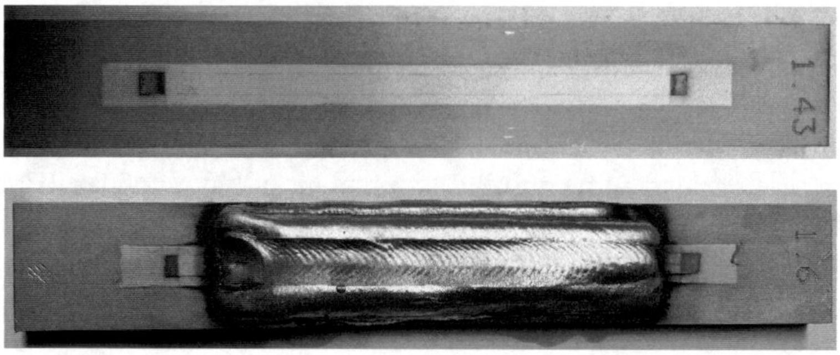

Figure 5. a, Sample after thermal spraying of alumina (white coloured) and copper tracks (orange) onto aluminium substate before the planned WAAM operation; b. Sample after integration of insulated conductor stack by aluminium WAAM layers.

Figure 6. Experimental clamping system with a test sample (without alumina insulation) exhibiting the contact pins used for both the electrical power supply as well as the resistance measurement.

applied but only for less than 1 s, on the other hand the so called long-time variant, by which the conductor tracks were heated up to 580 °C for up to 5 min applying lower ampere setting.

4.1 Short-time electrical processing

The so-called short-term variant aimed at the minimum of treatment time and energy consumption, respectively.. A welding power source (Fronius MagicWave 5000) was used to provide a current of up to 500 A. The electrical current was introduced directly into the conducting copper layer stack by pins contacting both ends of the conductor track (Figure 6). The experimental variations included the combination of different process durations (140 ms, 880 ms) with different current settings (225 A, 325 A, 500 A). Due to the fast heating, an exact temperature measurement was not available.

4.2 Long-time electrical processing

The so-called long-time experiments applied a welding power source (Fronius MagicWave 2000) limited to a current of max. 180 A. The experimental setup was similar to the setup in the short-time experiments with the main difference in the applicability of thermocouples, located under the aluminium substrate in order to measure the substrate's temperature, acting as the main process parameter. The chosen temperatures values were 250 °C, 350 °C and 580 °C, respectively. By applying a current of 180 A to the copper conductor track, the overall sample temperature increased by time. When the dedicated temperature was reached, the power supply ended.

4.3 Introduction of the novel treatment options into the process chain

While the standard heat treatment was applied only to samples with WAAM layers already deposited, the two electrical processing variants ("short time", "long time") were already applied to two subsets of samples in advance of the WAAM operations.

In the legend of Figure 7, the timing of the application of one of the two novel process variations to a certain subset is stated. The resistivity values of five subsets were measured at five stages; the processed results are shown as clusters along the x-axis in Figure 7. The first cluster of results is called "Incoming", showing all average values and related error bars derived from all intact samples after APPD processing before any further steps. The next cluster - called "Before WAAM" - includes two bars providing the results after "short time" or "long time" post-treatment, respectively. So far, all samples looked similar to the one shown in Figure 5.a. The next cluster in Figure 7 shows the specific resistance results "After WAAM", i.e. all intact samples are similar to the one shown in Figure 5.b, repeatedly showing also the data for samples without any post-processing. While the subset "Before WAAM long time" had to be withdrawn from further investigations due to damage to the contact areas (obviously resulting from thermo-mechanical stresses caused by fast thermal loading in combination with the local pressure from clamping), all remaining samples exhibited a reaction to the heat generated during the integration operations by WAAM. A significant drop in specific electrical resistance was observed compared to the data of incoming samples. The cluster "Before heat treatment" includes two subsets which were electrically treated after the WAAM integration. They were named "After WAAM short time" and "After WAAM long time". Only for the later one as well as the reference subset called "No treatment" (referring to the novel post-treatment) standard heat treatment was applied, the two bars clustered to the data set "After heat treatment".

5. Main results

The most important finding of these studies is that both a standard T6 heat treatment as well as a special electrical post-processing were improving the electrical conductivity of copper conductors integrated in an aluminium body significantly. Furthermore it was found that even the thermal energy of the WAAM process introduced into the conductor stack led to a significant drop in electrical resistivity to the level of the values after electrical treatment. In Figure 7, the resistivity results of five subsets are shown in clusters of bars, After the APPD processing, the measurement revealed values in the range of 0,05 to 0,075 Ω mm^2/m. Two subsets were treated by one of the two electrical treatments, leading to average values of 0,035 Ω mm^2/m. These numbers are at the same time at 70 - 50% of the level of the incoming samples but still around 200% of the standard properties of pure copper (0,017 Ω mm^2/m). Nevertheless, the potential of the novel electrical treatment was shown by this improvement.

After the integration by WAAM, all samples exhibited increased electrical conductivity – slightly less than the ones electrically treated after APPD showed but significantly better than without any treatment, altogether at a resistivity level of appr. 0,04 Ω mm^2/m.

The best values were achieved by a combination of WAAM integration, post-processing by the long-time variant of this novel electrical treatment plus a standard T6 heat treatment. Despite the fact that the "short-time" variant failed to produce enough comparable data available for evaluation, the level of the specific electrical resistance data of samples treated only by WAAM and T6 HT is very near to the best values. The two best average values are 0,025 and 0,027 Ω mm^2/m respectively, which are about 50-60% higher than the standard values of pure copper but much lower than the values derived from incoming samples in their as-deposited state.

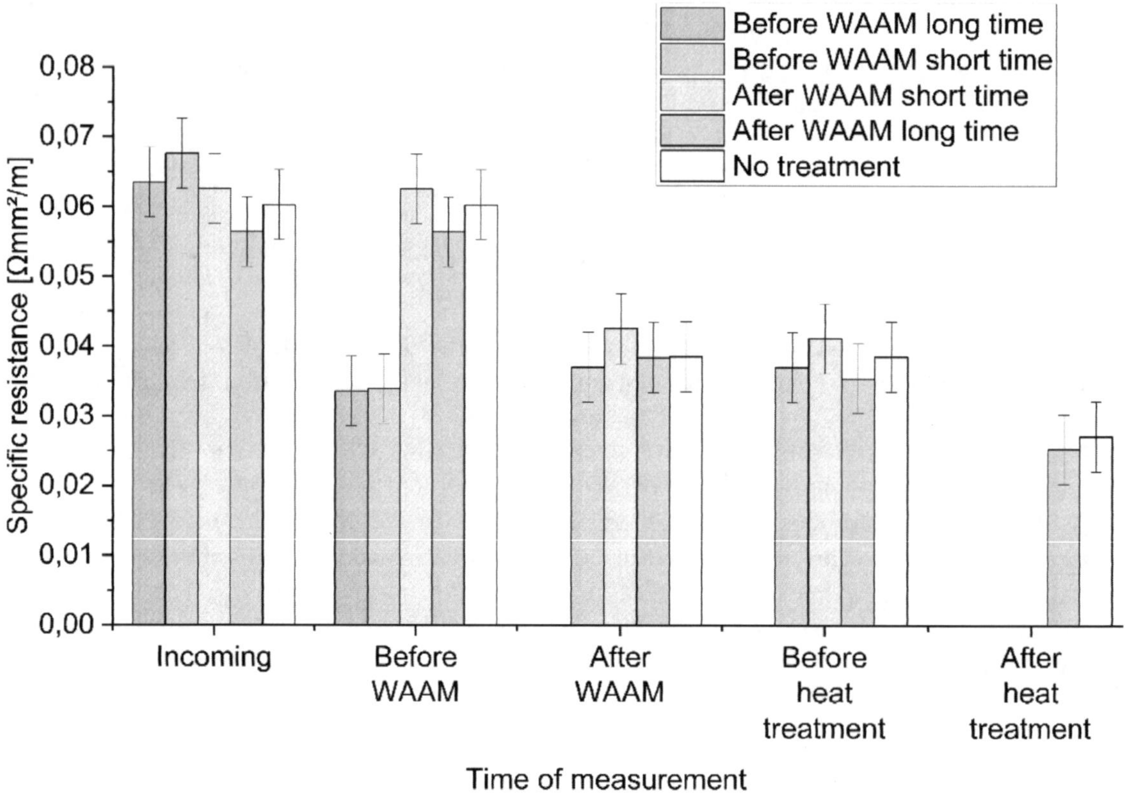

Figure 7. Specific electrical resistance of samples treated along the process chain with and without novel electrical post-processing [4, 8].

6. Conclusions

The novel method of electrical post-treatment has clearly shown the potential to increase the electrical conductivity of copper layers deposited by plasma spraying methods. For aluminium parts which are not applied to standard heat treatment - and when the conductivity improvement by the heat input of WAAM are considered to be not enough - the short-term electrical treatment is offering a very quick and energy-efficient process to reduce resistance by half in under one second by using a conventional power source. This effective procedure seems to have a high potential to be exploited with insulated copper conductors embedded into other substrate materials as well.

Since the thermal sprayed copper tracks are reacting to heat input with an improvement of the electrical conductivity, any T6 heat treated aluminium structures are expected to achieve very similar levels of improvement without additional electrical post-processing of the copper tracks. Hence, manufacturers of aluminium parts with integrated copper conductors have now the opportunity to select the post-processing method according to required properties and process chain in place.

Acknowledgements

The project MULTI-FUN has received funding from the European Union's H2020 research and innovation programme under Grant Agreement No. 862617. This publication reflects only the authors views. The European Commission and the European Health and Digital Executive Agency (HaDEA) are not responsible for any use that may be made of the information it contains.

References

[1] Gradinger R, 2023, Progress in multi-material design – Results from the HORIZON 2020 project MULTI-FUN, Metal Additive Manufacturing Conference 2023, October 17-19, 2023, Vienna, Austria
[2] Kordaß R, Schramm M, Gradinger R, Multi-material Functional Integration with Combined DED Additive Manufacturing Processes. ATZ Worldw 126, 42–45 (2024). https://doi.org/10.1007/s38311-024-1936-1
[3] Gradinger R, Schnall M, Hohm D, Hinterer A, Spalt S, 2022, Investigations into the robustness of additively manufactured, embedded electrical conductors, Metal Additive Manufacturing Conference 2022, September 26-28 , Graz, Austria
[4] Hohm D, 2023, Bachelor Thesis, Montanuniversität Leoben
[5] Cavaliere P, Sadeghi B, Shabani A, Spark Plasma Sintering: Process Fundamentals. In: Cavaliere, P. (eds) Spark Plasma Sintering of Materials. Springer Cham. 2019, https://doi.org/10.1007/978-3-030-05327-7_1
[6] www.inocon.at/wp-content/uploads/2023/08/ICN_Prospekt_InoCoat_EN.pdf, July 31 2024
[7] www.migal.co/en/products/aluminium-welding-wire/ma-6063-almg07sitib, July 31 2024
[8] Hohm D, Gradinger R, 2023, Electrical treatment of thermal sprayed copper layers to improve the electrical conductivity, Proc. of Metal Additive Manufacturing Conference 2023, pp. 59-68

Effect of welding gas mixtures on weld material and bead geometry

Michael Unger*[1], José L. Neves[1], Martin Schnall[1]

[1] LKR Light Metals Technologies, AIT Austrian Institute of Technology, 5282 Ranshofen, Austria

*E-mail: michael.unger@ait.ac.at

Abstract. Shielding gas is an essential parameter in welding processes, serves to shield the molten material from the atmosphere and external contaminations while providing a stable electric-arc pathway during the process, thus influencing the quality of the weld bead and several other aspects of the process. Therefore, any welding-based process, such as wire-arc additive manufacturing, could benefit from shielding gas mixture optimisation to selectively improve the processing and possibly the final mechanical properties of the welded material. In this study welding of EN AW 5183 aluminium (single weld bead and wire-arc additive manufacturing build-up) was conducted using several gas mixtures and their influence on final geometry, defects, and microstructure measured. The gas mixtures used were a combination of varying amounts of carbon dioxide, nitrogen, and oxygen. These were added to commercially available argon gas. It was observed that: CO_2 additions led to an unfavourable bonding defect and increased mean grain size; nitrogen additions caused a decrease in mean grain size, decreased pore count while maintaining a similar pore volume fraction to the reference argon mixture samples; oxygen additions showed slight reduction of mean grain size, decreased pore count and decreased pore volume fraction. These results further showcase that varying shielding gas mixtures influence the final material properties and should be considered as an additional improvement route for conventional welding and additive manufacturing.

1. Introduction

First invented, in 1986, additive manufacturing (AM) technology, has received an ever-growing interest from academia and industry due to its potential in several sectors. This potential can be characterized by an increase in design freedom, reduction of material consumption and reduced energy use [1]–[3].

AM encompasses a wide range of processes categories. One of the most popular process categories of AM, is wire-arc directed energy deposition (waDED). This process can be used for the production of large near net shape structures with complex geometries and can also be applied to repair use-cases. The process consists of locally feeding a wire feedstock material, which is molten selectively with the use of an external heat source. For this process category, the heat sources are conventional welding technologies based on electric arcs, such as gas metal arc welding (GMAW), gas tungsten arc welding (GTAW), and plasma arc welding (PAW) [1], [4].

Due to its susceptibility to contamination from the atmosphere, the process needs to be locally shielded by using shielding gases. Besides the shielding purpose, shielding gas also has a significant effect on the process as it can influence arc stability (voltage), deposition rate, heat transfer, weld penetration, bead form and contributes to the composition of the weld material and mechanical properties. One mechanic on how the used gas influences this is that it needs to be dissociated and has a different ionization energy and thermal conductivity that are proven to have the mentioned effect [5]–[8].

This means that selecting the correct gas mixture for shielding gas in waDED can ensure optimal conditions for part building, possibly benefiting process efficiency, accuracy, local alloying possibilities, and ultimately, final properties of the material. Therefore, in this study, different additions of nitrogen (N_2), oxygen (O_2) and carbon dioxide (CO_2) were added to commercially available argon gas mixture (Ar), commonly used in welding. These gas mixtures were then used as the shielding gas for depositions of the filler wire, EN AW 5183, and their influences on the process and final properties measured [6], [8]–[11].

This study aims to improve the understanding of the effects and potential of dopped gas mixtures use in the waDED processes.

2. Experimental

2.1. DoE and trials

The reference gas mixture used was commercially pure Ar 5.0 with an 99.99% purity. To this reference mixture varying ranges of the mentioned doping gases, were added. The addition ranges and increments are as follow: N_2 additions were done up to 5% of the total gas mixture, using 2.5% increments; CO_2 additions were done up to 5% of the total gas mixture, using 2.5% increments; O_2 additions were done up to 0.03% of the total gas mixture, using 0.015% increments. With the help of the design of experiment (DOE) method a sample plan was established as can be seen in **Table 1** [12]. The filler metal used was EN AW 5183 from MIGAL.CO GmbH, with a diameter of 1.2 mm.

The experiments are divided in two stages, initially a single weld bead is deposited on a substrate and preliminary results analysed before proceeding with a multi-pass build-up of material. The two stages were conducted using two different motion systems and two different power sources. These changes are stated in the following sub-chapters. The change in motion and heat source equipment was only due to laboratory and resource capabilities.

Deposited samples were characterized by taking microsections of the weld beads. These were embedded in resin and then polished and investigated using optical light microscopy, (Olympus BX60) with whom geometry effects and porosity can be measured and quantified. The analysis of porosity and the measurements of grainsize were done with the help of the free software "Fiji ImageJ" [13].

Table 1. Sample plan of the single weld bead samples, generated by application of a DOE method.

Sample name	N_2 in %	CO_2 in %	O_2 in %	Ar in %
V1_N_2_2.5_CO_2_2.5_O_2_0.015	2.5	2.5	0.015	94.985
V2_ CO_2_5	0	5	0	95
V3_N_2_5_CO_2_5_O_2_0.03	5	5	0.03	89.97
V4_N_2_5_CO_2_5	5	5	0	90
V5_N_2_5_O_2_0.03	5	0	0.03	94.97
V6_CO_2_5_O_2_0.03	0	5	0.03	94.97
V7_ref_Ar_100	0	0	0	100
V8_N_2_5	5	0	0	95
V9_O_2_0.03	0	0	0.03	99.97

Following the geometry and porosity measurements the samples were chemically etched using the Barker´s agent revealing the grain structure of the samples, allowing for grain measurements via the linear interception method according to ASTM E112–13.

2.2. Single weld bead

For the deposition of the single weld bead, using EN AW 5183 as filler wire and base plates of EN AW 6060 alloy with a length of 200mm, a width of 100 mm and a thickness of 10 mm. The samples were manufactured using the KUKA KR300 R2700-2-FLR robot motion system coupled with a Fronius power unit TPS 4000 CMT and equipped with a Fronius Robacta 5000 welding torch.

To ensure the correct gas mixture compositions for each experiment, a GA-MI-6.2 mixing unit manufactured by the company Gasyco GmbH, was used. The gas mixing unit and gas pipes were flushed continuously for 15 minutes to avoid contamination (humidity and oxygen). The baseplates were brushed with a wire brush and cleaned with isopropanol prior to depositions. The single weld beads are exemplified in **Figure 1**, where the gas mixture used was the reference gas mixture.

IOP Conf. Series: Materials Science and Engineering **1315** (2024) 012007
doi:10.1088/1757-899X/1315/1/012007

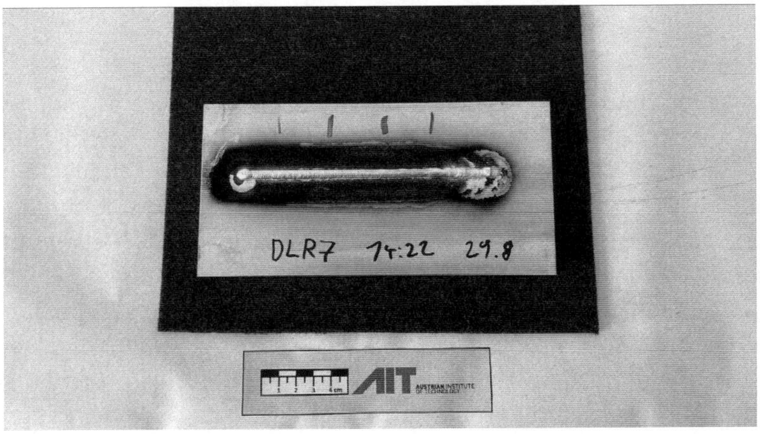

Figure 1. Reference sample of the bead on plate trials.

2.3. Manufacturing wall geometry

As previously stated, the robot-workplace was switched to a different motion system, ABB IRC 8 axis 4600 and the power source was a Fronius CMT Advanced 4000 R. Lastly, the substrate material used was also changed due to a lack of material, however the measurements near the first layers were not considered hence comparability was not compromised. The geometry of the samples are rectangular, as depicted in **Figure 2**, with rounded corners for better continuity of the weld. The Length of the samples are 150 mm, and the width 40mm.

Figure 2. Reference sample of the wall geometry samples

3. Results

3.1. Single weld bead

During welding soot buildup is expected at the flanks of the weld. This can be observed for the reference depicted in **Figure 1**. The addition of O_2 resulted in a similar appearance and soot buildup as the reference sample. Samples with N_2 and CO_2 showed an increase of soot build-up at the flanks and on the bead surface. Furthermore, the mixtures containing CO_2, upon visual inspection, showed increased surface texture (groves and cavities). N_2 samples, containing no CO_2, showed bronze discolouring, and had decreased soot build-up on the bead itself.

The microsections where taken from three different sample positions. These showed different bonding behavior and other contact angles to the substrate plate. The CO_2 samples appeared to have a bonding defect at the surface level of the substrate plate. This can be seen in the top row of the selected samples in **Figure 3** where two mixtures containing CO_2 are depicted.

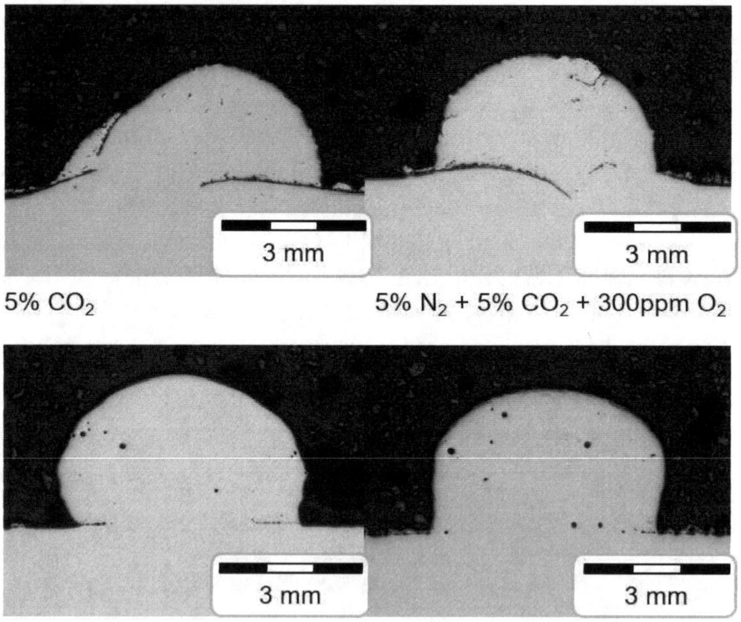

5% CO_2 5% N_2 + 5% CO_2 + 300ppm O_2

100% Ar (ref.) 5% N_2

Figure 3: Representative selection of the microsection from bead on plate samples.

Geometric measurements showed a reduction in height for all samples while the width followed no clear trend. The height of the reference sample was, 3.27 ± 0.17mm, which is the maximum in all samples while the lowest sample with 5% N_2 5% CO_2 and 300 ppm O_2 was measured to be 2.64 ± 0.06mm. The width measurements are as follows: the reference sample has a width of 5.17 ± 0.36mm, while the sample 5% CO_2 and 300ppm O_2 with 6.57 ± 0.16mm representing the highest value and the lowest measurement is reached by 5% N_2 and 300ppm O_2, measured with 4.77 ± 0.20 mm (**Figure 4**).

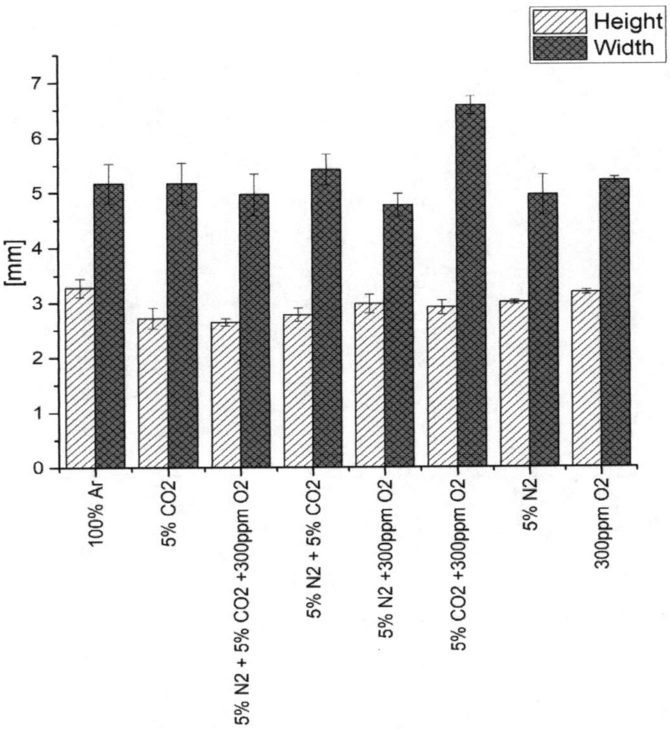

Figure 4: Mean values of geometry measurements for bead on plate samples.

Porosity analysis shows the cross-section area of the pores and the average pore count per microsection. The average pore count of the reference was 47 pores, a count that the samples with N2 and the sample with N2 + 02 reached a lower value of. The other samples had approximately 20,40 or 80 pores more than the reference with the maximum reached by sample 5% CO_2+300ppm O_2 that had 133 pores on average that also had one of the lowest standard deviations with 4.64 pores. Those samples and measurements are shown on the right in **Figure 5**.

The cross-section area of the pores in the reference was 0.59%±0.13% which is in the range of three other samples whose values reach from 0.53% to 0.64% while four other samples range from 0.09% to 0.27%. this can be seen in **Figure 5** on the left side.

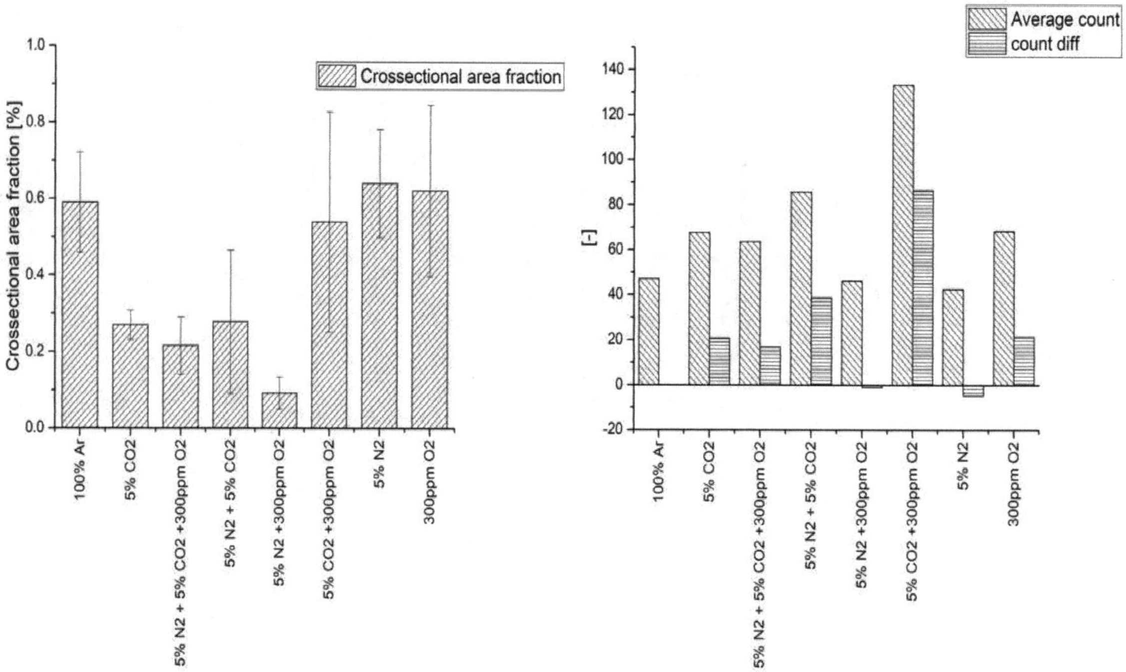

Figure 5: Results of porosity measurements of the single bead depositions. Mean pore area fraction shown on the right and pore count with difference to reference sample on the left.

Grainsize measurements were conducted according to the mean linear intercept method described in the ASTM 112-13 standardization. Results are displayed, in **Figure 6**, according to measurement direction, vertical and horizontal. The vertical direction for the samples is the bead height while the horizontal direction is parallel to the bead width. Quantitative values are given in **Table 2**.

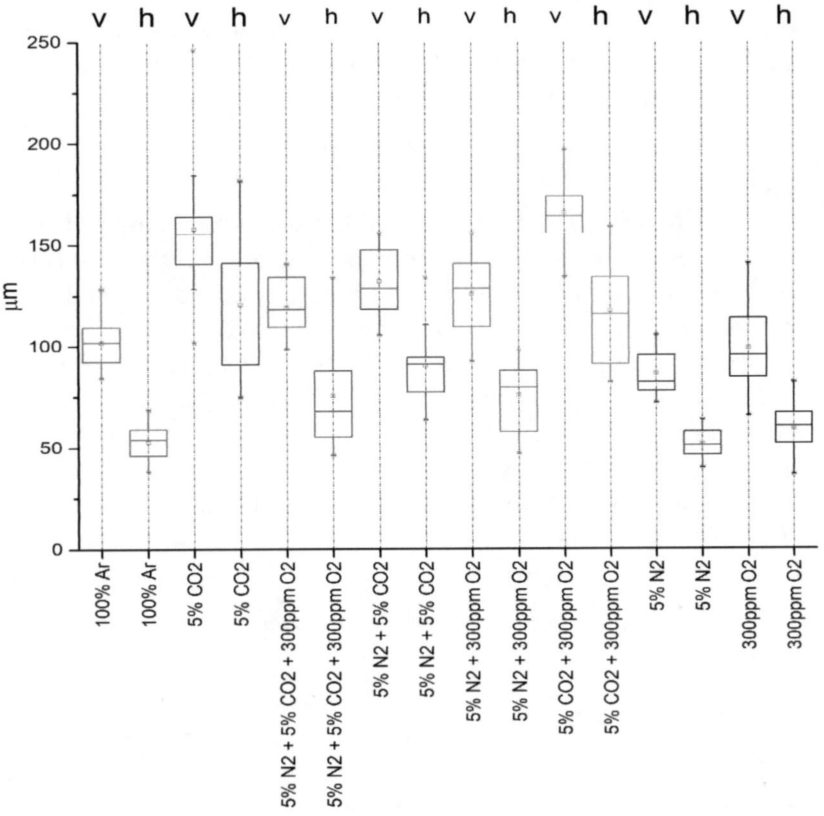

Figure 6: Grainsize of single weld bead trials in vertical and horizontal direction.

Table 2: Table of median grainsize in vertical and horizontal direction.

Gas mixture	Vertical [µm]	Horizontal [µm]
100% Ar	101.72	54.04
5% CO_2	155.26	141.11
5% N_2 5% CO_2 300ppm O_2	118	67.75
5% N_2 5% CO_2	128.26	90.71
5% N_2 300ppm O_2	128.26	79.38
5% CO_2 300ppm O_2	163.89	115.45
5% N_2	81.94	50.8
300 ppm O_2	95.16	60.48

The reference sample has a median grainsize of 101.7µm in the vertical and 54 µm in horizontal direction. The 300 ppm O_2 sample, has a grainsize 95.16 µm and 60.48 µm, in the vertical and horizontal directions, respectively. The N_2 sample has grains of the size 81.94 µm and 50.8 µm, in the vertical and horizontal directions, respectively.

3.2. Wall geometry

For wall geometry manufacturing it was decided to exclude the samples that contain CO_2 in the mixture due to the mentioned bonding defect mentioned in subsection 3.1 Single weld bead.

While welding with the remaining five gas mixtures appeared greater soot build-up with advancing progression in buildup and when using N_2 in the gas mixture. Furthermore, the surface showed groves along the building layer and waves or distortions on the general wall surface. In addition, a bronze surface coloring could be observed when mixtures containing N_2 are used.

In **Figure 8** a representative section for the measurement method for wall width is shown and in **Figure 7** the minimum, mean and maximum values are plotted. The mean width is decreasing for all samples when compared to the reference sample (4.89mm).

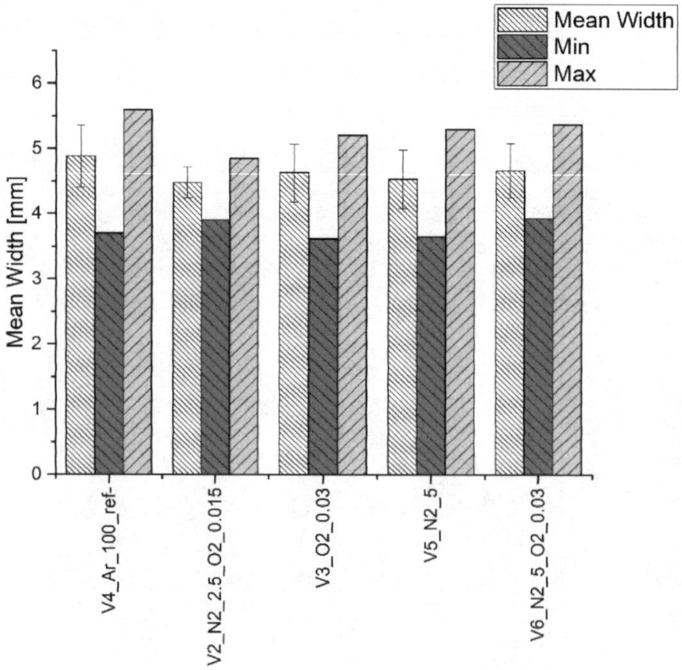

Figure 7: Measurements of wall sample width.

IOP Conf. Series: Materials Science and Engineering **1315** (2024) 012007 doi:10.1088/1757-899X/1315/1/012007

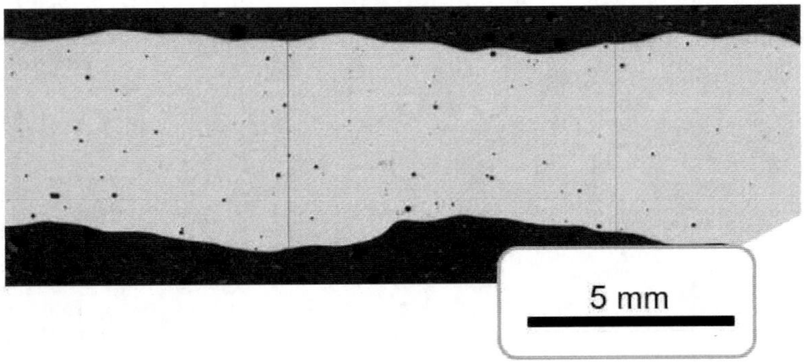

Figure 8: Representative pictures of wall microsection with measurement marks.

In **Figure 9** a) the measured pore count of each sample is given. The pore count decreases for samples with added gases compared to the reference gas mixture. However, pore density is not directly correlated to the volume fraction of pores, as shown **Figure 9**b), the pore volume fraction remains relatively similar between samples as the mean values are in the range of 0.18% ± 0.05% to 0.24% ± 0.1%.

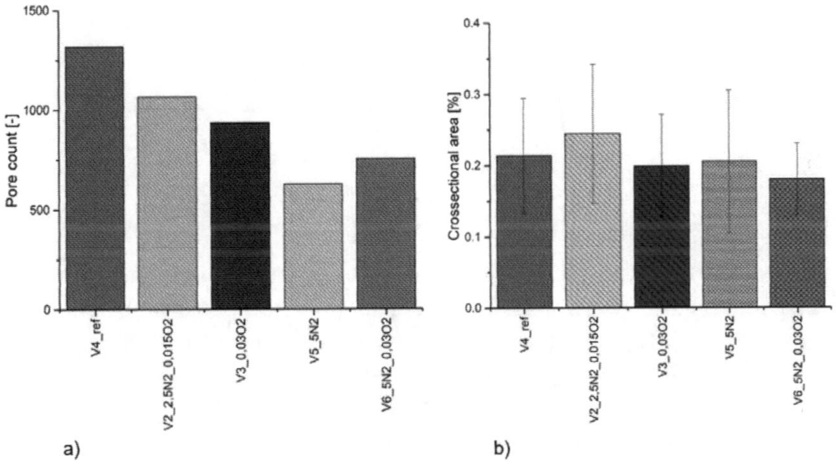

Figure 9: Porosity analysis of wall samples a) shows pore count while b) pore volume fraction.

Grainsize measurements showed that the biggest grains are found in the reference samples and the smallest grains in samples with N_2+O_2 in the gas, shown in **Figure 10**.

The reference sample has a mean grainsize of 56.13 ± 8.20 μm in the vertical direction (thickness direction of the wall) and 67.85 ± 7.19 μm in the horizontal direction (building direction) while the smallest is reached at 41.46 ± 5.60 μm in vertical and 47.58 ± 5.01 μm by the sample with 5%N_2 and 0.03%O_2. The samples with 300 ppm O_2, showcase grains that are slightly refined compared to the reference samples.

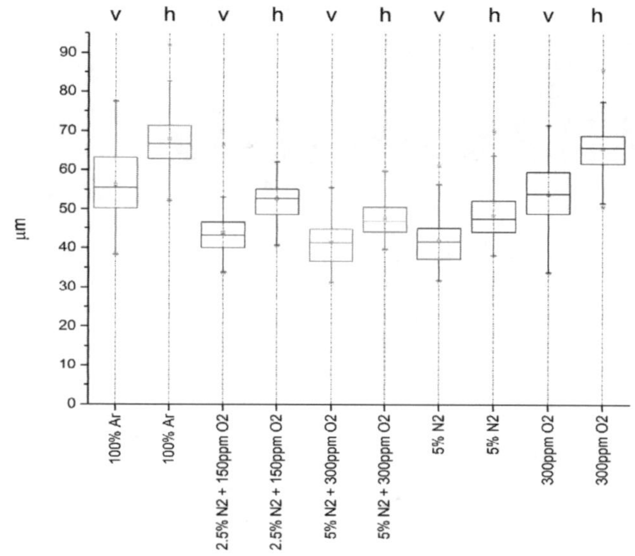

Figure 10: Grainsize boxplot of wall samples with different gas mixtures.

4. Discussion

The differences of the surface and soot build up described in the results for the single weld bead samples with CO_2 is partially explained for steels in the study of Mvola and Kah [9]. It was described that the dissociation of CO_2 in the gas led to oxidation of the surface during the processing, and at the same time this results in a change of energy potential thus increasing current density and ultimately causing more spatter and soot. Another possibility for the soot build-up is approached by Menzel [14], who described that increased O_2 and CO_2 contents are raising burnup of alloy elements. Additionally, in the paper of Li *et al.*[8] it is shown that the difference in droplet transfer when using Ar and N_2 ultimately resulted in higher spatter in case of nitrogen.

The bead geometry is affected by many parameters simultaneously or as J.M. Kuk et al.[6] states it is a very complex relationship of all the welding parameters. One important measure for the wall samples is the minimum measured width because it represents the possible end geometry.[15]–[17] It is observed that the mean width of the wall samples is reduced compared to the reference while the single weld bead samples show no clear trend in width in correlation with used shielding gas, as **Figure 4** shows. The measurements for the min wall width (**Figure 7**) range from 3.62mm to 3.93mm. The min wall width of the reference is 3.71 mm.

The height of the single weld bead samples is lower for all tested gas mixtures compared to the reference, see **Figure 4**. The sample 5% N_2 5% CO_2 and 300 ppm O_2 had the minimum measured height of all samples with a reduction of 19.27%. The samples with CO_2 in the mixture are the four samples with the lowest height, this can be explained by the increased amount of available oxygen in the plasma that is reported by Schafraski, da Cunha and Bohórquez [18] that decreases the surface tension of the melt. This causes the height to decrease. According to Li et al [8] it is expected to increase due to an decrease of heat input when dissociation of the gas appears to get ionized for the arc.

For the porosity for both the bead on plate and wall samples it can be said that there is no visible correlation between pore count and pore area fraction. This can be seen in **Figure 9** where the trend of reducing pore count is not reflected by the pore volume fraction as expected, meaning that despite reduced pore count, these increased in size. The highest pore count in the wall samples is measured in the reference sample, while the other mixtures have a reduced count, this trend seems to also be present for the single weld bead samples.
According to Kah and Martikainen [17], greater fluidity of the melt pool can give hydrogen, or other entrapped gases, more time to escape and therefore reduce pore count. This can be achieved by increasing the heat input, which is an effect they also attribute to nitrogen, which contradicts the work of Li et al.[8] as stated previously. Aldalur, Suárez and Veiga [19] reports that a lower droplet transfer rate is increasing porosity in samples. As Li et al. [8] reported in their work, an increase in droplet transfer appears under nitrogen, which explains the decrease of pores in this work via the same mechanism.

Another aspect of the material that is affected by heat input is the grain size. The boxplot in **Figure 6** shows that the sample with CO_2 and CO_2 + O_2 have a greater grain size than the other single bead samples and combinations. N_2 in combination with CO_2 and/or O_2 grain size remains constant but still increased compared to the reference. It is reported that CO_2 is increasing the heat input in welding and therefore increasing the growth time. In the work of S. Li, et al.[8] the grain size reducing effect of N_2 is also present and can be explained by the decrease of heat input that is caused by the energy needed to dissociate the molecule [20]. One further explanation for the grain refinement is that in the arc N_2 dissociates and is now atomically in the plasma which enables the gas to interact with the molten aluminium to form aluminium nitride (AlN)[21]. Nitrides formed can act as a grain refiner, which is known in other materials besides aluminium such as steel and titanium[20], [22], [23].

5. Conclusion

The changes in Process and resulting material when different gas mixtures are used for welding and waDED have been investigated and can be summarized as follows.

1. CO_2 promotes bonding defect when used in shielding gases.
2. Nitrogen reduces porosity in bead on plate welds and wall structures manufactured with waDED.
3. Nitrogen and small amount of O_2 refines the grainsize of the material.
4. CO_2 appears to be increasing the grainsize.

Acknowledgements

The Autor wants to thank Stephan Ucsnik for the preparation beforehand, his consultancy throughout and managing the project as consortium lead, C. Schneider-Broeskamp and A. Birgmann for their welding expertise and time for sample manufacturing. Furthermore, Sebastian Zehetner from Linde Gas GmbH for providing the mixing unit and the gases to perform the trials and further all involved staff members of the LKR for their assistance and expertise.

The authors would like to thank the ministry of "Climate Action, Environment, Energy, Mobility, Innovation and Technology" (BMK), the ministry of "Digital and Economic Affairs" (BMDW), the Austrian Funding Agency (FFG), as well as the four federal funding agencies Amt der Oberösterreichischen Landesregierung; Steirische Wirtschaftsförderungsgesellschaft m.b.H.; Amt der Niederösterreichischen Landesregierung; Wirtschaftsagentur Wien, ein Fonds der Stadt Wien for funding project "We3D" (grant agreement no. 886184) in the framework of the 8th COMET call.

Further thanks to the industry partners for their financial and research contributions in the five conducted multi-firm projects.

References

[1] K. Kanishka and B. Acherjee, "Revolutionizing manufacturing: A comprehensive overview of additive manufacturing processes, materials, developments, and challenges," *J. Manuf. Process.*, vol. 107, pp. 574–619, 2023, doi: 10.1016/j.jmapro.2023.10.024.

[2] S. H. Huang, P. Liu, A. Mokasdar, and L. Hou, "Additive manufacturing and its societal impact: a literature review," *Int. J. Adv. Manuf. Technol.*, 2013, doi: 10.1007/s00170-012-4558-5.

[3] B. Wu *et al.*, "A review of the wire arc additive manufacturing of metals: properties, defects and quality improvement," *J. Manuf. Process.*, vol. 35, pp. 127–139, Oct. 2018, doi: 10.1016/J.JMAPRO.2018.08.001.

[4] M. Köhler, S. Fiebig, J. Hensel, and K. Dilger, "metals Wire and Arc Additive Manufacturing of Aluminum Components," 2019, doi: 10.3390/met9050608.

[5] T. Sato, "Influence of shielding gases on quality and efficiency in gas shielded arc welding," *Weld. Int.*, vol. 15, no. 8, pp. 616–619, 2001, doi: 10.1080/09507110109549413.

[6] J. M. Kuk, K. C. Jang, D. G. Lee, and I. S. Kim, "Effects of temperature and shielding gas mixture on fatigue life of 5083 aluminum alloy," *J. Mater. Process. Technol.*, pp. 1408–1414, 2004, doi: 10.1016/j.jmatprotec.2004.04.117.

[7] R. D. Ardika, T. Triyono, N. Muhayat, and Triyono, "A review porosity in aluminum welding," *Procedia Struct. Integr.*, vol. 33, no. C, pp. 171–180, Jan. 2021, doi: 10.1016/J.PROSTR.2021.10.021.

[8] S. Li *et al.*, "Comparative study on the microstructures and properties of wire+arc additively manufactured 5356 aluminium alloy with argon and nitrogen as the shielding gas," *Addit. Manuf.*, vol. 34, p. 101206, Aug. 2020, doi: 10.1016/J.ADDMA.2020.101206.

[9] B. Mvola and P. Kah, "Effects of shielding gas control: welded joint properties in GMAW process optimization," *Int. J. Adv. Manuf. Technol.*, vol. 88, no. 9–12, pp. 2369–2387, Feb. 2017, doi: 10.1007/S00170-016-8936-2/METRICS.

[10] B. G. Chung, S. Rhee, and C. H. Lee, "The effect of shielding gas types on CO2 laser tailored blank weldability of low carbon automotive galvanized steel," *Mater. Sci. Eng. A*, vol. 272, no. 2, pp. 357–362, Nov. 1999, doi: 10.1016/S0921-5093(99)00499-2.

[11] C. Matz and E. Miklos, "Method of Arc-Joining," US2007/0045237 A1, Mar. 01, 2007

[12] K. Siebertz, D. van Bebber, and T. Hochkirchen, "Statistische Versuchsplanung," 2017, doi: 10.1007/978-3-662-55743-3.

[13] J. Schindelin *et al.*, "Fiji: An open-source platform for biological-image analysis," *Nat. Methods*, vol. 9, no. 7, pp. 676–682, Jul. 2012, doi: 10.1038/NMETH.2019.

[14] M. Menzel, "The influence of individual components of an industrial gas mixture on the welding

process and the properties of welded joints," *Weld. Int.*, vol. 17, no. 4, pp. 262–264, 2003, doi: 10.1533/WINT.2003.3111.

[15] H. Wang, W. Jiang, J. Ouyang, and R. Kovacevic, "Rapid prototyping of 4043 Al-alloy parts by VP-GTAW," *J. Mater. Process. Technol.*, vol. 148, no. 1, pp. 93–102, May 2004, doi: 10.1016/J.JMATPROTEC.2004.01.058.

[16] T. Klein and M. Schnall, "Control of macro-/microstructure and mechanical properties of a wire-arc additive manufactured aluminum alloy," *Int. J. Adv. Manuf. Technol.*, vol. 108, no. 1–2, pp. 235–244, May 2020, doi: 10.1007/S00170-020-05396-6/FIGURES/6.

[17] P. Kah and J. Martikainen, "Influence of shielding gases in the welding of metals," *Int. J. Adv. Manuf. Technol.*, vol. 64, no. 9–12, pp. 1411–1421, Feb. 2013, doi: 10.1007/S00170-012-4111-6.

[18] L. L. Schafranski, T. V. da Cunha, and C. E. N. Bohórquez, "Benefits from H2 and CO2 additions in argon gas mixtures in GMAW," *J. Mater. Process. Technol.*, vol. 249, pp. 158–166, Nov. 2017, doi: 10.1016/J.JMATPROTEC.2017.05.030.

[19] E. Aldalur, A. Suárez, and F. Veiga, "Metal transfer modes for Wire Arc Additive Manufacturing Al-Mg alloys: Influence of heat input in microstructure and porosity," *J. Mater. Process. Technol.*, vol. 297, p. 117271, Nov. 2021, doi: 10.1016/J.JMATPROTEC.2021.117271.

[20] I. Boiko and D. Avisans, "STUDY OF SHIELDING GASES FOR MAG WELDING," 2013.

[21] Q. Hou, R. Mutharasan, and M. Koczak, "A Feasibility of aluminium nitride formation in aluminum alloys," *Mater. Sci. Eng.*, vol. 195, pp. 121–129, 1995.

[22] M. Bermingham, D. StJohn, M. Easton, L. Yuan, and M. Dargusch, "Revealing the Mechanisms of Grain Nucleation and Formation During Additive Manufacturing," *JOM*, vol. 72, no. 3, pp. 1065–1073, Mar. 2020, doi: 10.1007/S11837-020-04019-5/FIGURES/7.

[23] T. Yuan, X. Ren, S. Chen, and X. Jiang, "Grain refinement and property improvements of Al–Zn–Mg–Cu alloy by heterogeneous particle addition during wire and arc additive manufacturing," *J. Mater. Res. Technol.*, vol. 16, pp. 824–839, Jan. 2022, doi: 10.1016/J.JMRT.2021.12.049.

IOP Conf. Series: Materials Science and Engineering **1315** (2024) 012008 doi:10.1088/1757-899X/1315/1/012008

Overview on aluminium alloys as sinks for end-of-life vehicle scrap

Stefan Pogatscher[1] and Sebastian Samberger[1]

[1], Chair of Nonferrous Metallurgy, Montanuniversitaet Leoben,
Franz-Josef Straße 18, 8700 Leoben, Austria

*E-mail: stefan.pogatscher@unileoben.ac.at

Abstract. A fundamental principle in metallurgy is that the higher the purity of metals and alloys, the more favourable their properties will be. However, as the recycling of materials in production becomes increasingly significant, the levels of impurities are also on the rise. In the case of aluminium, the consequences can be detrimental due to the low solubility of most elements in this metal, which leads to the formation of brittle intermetallic phases (IMPs). Moreover, once impurities have entered aluminium, it is difficult to remove them. In 2017, almost 100 million cars were produced worldwide. Historically, vehicle design prioritised performance, resulting in a multi-material mix to utilise the best materials for each application. This included over 40 different wrought and cast aluminium alloys, Cu-based materials for electrics, and steels for high-strength applications. In the recycling of end-of-life vehicles (ELVs), high purity wrought Al alloys are today down-cycled to low purity cast engine blocks. However, recent advancements show that the drawback of increase IMP-fractions can be turned into benefits through the strategic design of heterostructured alloys. A first successful alloy example from this approach enables interesting forming properties, previously only found in 5xxx series wrought aluminium alloys, in combination with a matrix composition and age-hardening potential known from 6xxx series wrought aluminium alloys. A second examples reviews compositions directly resulting from ELV scrap. By manipulating IMPs it is feasible to create heterostructures with an interesting balance of strength and ductility. These approaches challenge traditional views, allowing for a greater volume fraction of intermetallic phases. Understanding the formation and role of intermetallic particles is crucial. This work gives an overview to the current problem and the state of the art and addressed the potential of upcycled aluminium alloys that tolerate high impurity levels by using intermetallic phases as impurity sinks.

1. Introduction

Sustainable consumption and production, as called for in the Sustainable Development Goals of the United Nations, require substantial advances to be made in the efficient use of natural resources and recycling of waste. In addition to the objective of reducing the negative environmental impact of producing materials, recycling is particularly important in geographical regions where materials are not extracted in large quantities from natural resources but are used very extensively, as for example in Europe. There, recycling is of paramount importance if economies are to be less dependent on geopolitical constraints. This is illustrated in **Figure 1**, which shows the development of primary aluminium production from natural resources from

2000 to 2020 in various regions around the world. In global terms, Europe's relative importance is clearly declining [1]. This does not refer to the absolute amount produced, but while only 5% of primary aluminium was produced in Europe in 2020, 19% of all flat rolled aluminium products are still consumed in Europe [2]. All of this illustrates the potential future importance of recycling.

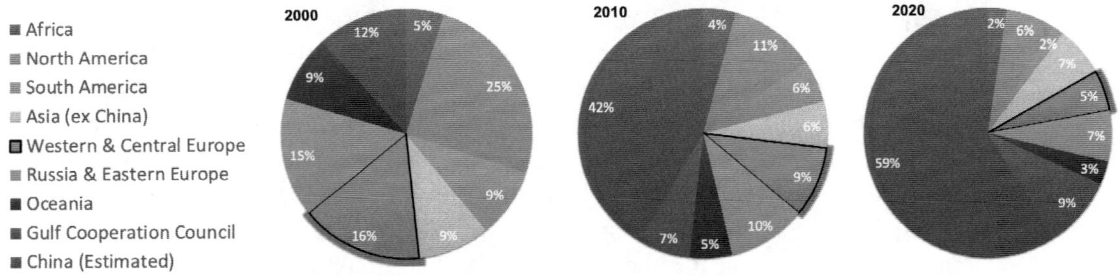

Figure 1. Proportion of primary aluminium production by region in 2000, 2010 and 2020 (data taken from the International Al Institute [1]).

While it is clear that sorting technology is also constantly improving, scrap that originates from several alloy types and different materials inevitably means an accumulation of impurities in recycled materials. This is highly problematic for Al, because its ignoble character hampers the metallurgical feasibility of chemical melt refinement for the vast majority of impurity elements [3, 4]. In principle, the purer metals and alloys are, the better are their properties. In Al this is particularly so because the solubility of most elements in Al is high in the liquid, but very limited in the solid state, with the result that brittle intermetallic phases (see section 2.1) form. [5]

Today more than twice as much Al is produced from natural resources than through recycling from scrap. For this reason, dilution with pure Al is still a possible, if non-sustainable, global strategy for mitigating impurity accumulation. However, this strategy will be hampered dramatically in the near future, for four potential reasons:

1. The share of scrap will rise significantly.
2. Wrought Al alloys which rely on primary Al are increasingly being used in vehicles.
3. Today's major scrap sinks – cast alloys for combustion engines – may decrease in their importance.
4. Many classes of Al alloys are compositionally incompatible.

In the following, we explain why these issues are of great importance from an economic and ecological point of view.

1. The trend is that the amount of scrap will significantly rise in Europe [6].
2. The increasing importance of high-performance wrought alloys in vehicles is also well documented [7–9], and is leading to rapid growth in the amount of scrap available from automotive wrought alloys [10].
3. By far the most important sinks for scrap today are cars' cast combustion engines, and rendering the automobile industry the largest consumer of recycled Al due to the large amount of cast alloys it deploys [9]. The AlSi6Cu3.5Zn and AlSi9Cu3.5FeZn alloys are most often used [11]. These cast alloys, which can absorb almost all wrought alloys [12], will drastically decline in importance over the next decade due to the rise of electric vehicles.

4. In contrast, today's wrought alloys represent roughly 2/3 of all the Al produced. They have narrow and often low compositional limits [2, 11] and are unable to absorb mixed alloy scrap streams, and therefore require pure primary Al [9, 13, 14]. For example, over 40 different, often compositionally incompatible Al alloys, together with steels and copper, are used in cars [9]. Such multi-metal design reduces the potential for recycling due to higher enrichment with detrimental impurities [15–17]: huge improvements in sorting technology would be necessary, and shredder-based recycling techniques have limits when confronted with multi-material products featuring different welding and joining techniques [18]. This situation will prospectively lead to an excess of wrought alloy scrap [8, 9] – even if we leave out the cast alloys from combustion engines which will arrive for recycling in the near future and which will possibly no longer be needed when electric vehicles predominate.

In view of the fact that Al has a global annual turnover of around USD 150 billion (in 2020 [19]), it is clear that the above constraints will have massive economic consequences. However, their ecological impact is also serious: the primary production of Al requires about 2% of the world's total energy consumption. It also generates significant greenhouse gas (GHG) emissions (2–3% of annual GHG worldwide) [20]. Recycling could in principle reduce both figures by 95% [21] – were it not for the four factors above. [22]

2. Current status

2.1 Detrimental effects of tramp elements in today's Al alloys

A longish list of problematic tramp elements are introduced by recycling, including Fe, Mg, Si, Mn, Ni, Pb, Cr, Cu, V and Zn [23]. Their challenging nature arises from today's high commercial alloy standards and from increasing requirements in the areas of alloy properties. The most advanced alloys in properties are also tolerate the lowest tramp element content today. One main reason for this is that the solubility of many elements in Al is very limited and the tendency to form intermetallic phases (IMPs) is high. These IMPs are usually undesirable, as they often cause properties to degrade. For example, minor fractions of IMPs containing Fe appear as brittle needles or plates (e.g. **Figure 2**), and these cause crack initiation [24, 25], deteriorate ductility and bendability [26], hamper fracture toughness [27] and a smooth surface finish [28]. In wrought alloys also Si is often considered critical due to the formation of Mg_2Si. Highly sensitive alloys include the 7xxx (AlZnMg) and 5xxx (AlMg) series of wrought alloys, because at high Mg content primary Mg_2Si IMPs are immediately formed [24]. This cases fracture at at lower strains than IMPs containing Fe [29, 30] such as plate-like β-Al_5FeSi IMP [31].

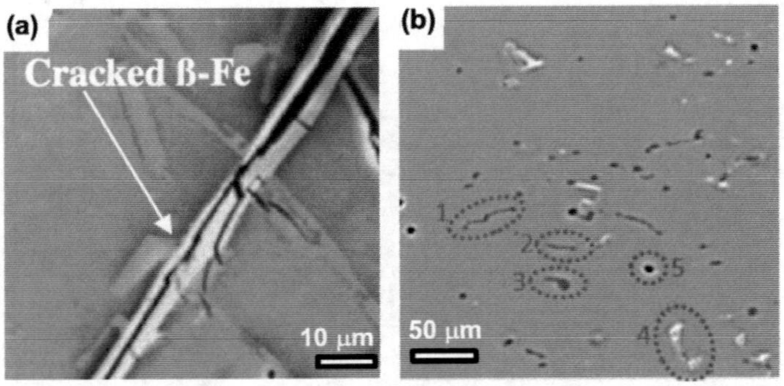

Figure 2. Examples of detrimental IMPs in commercial Al-alloys. (a) Cracked plate-like Fe containing β-particle [32]. (b) Initiation of damage by the decohesion of Mg_2Si IMP (1) in a 6xxx alloy [30].

Historically, the main method for mitigating the detrimental effect of IMPs in wrought aluminium alloys was to keep the content of primary IMP-forming elements low. This has been recognized as a general problem for future recycling [20], and an increasing number of researchers are trying to allay the effects of increased amounts of detrimental impurity elements (mainly focused on Fe) [22].

2.2 Known strategies for mitigating the negative effects of IMPs

Because casting alloys are still the main material in recycling, efforts have already been made there to mitigate the adverse effects of IMPs. Of the greatest interest have been Fe-containing IMPs, the formation of which was even studied on the International Space Station ISS [33]. The occurrence of detrimental morphologies in IMPs containing Fe can be mitigated by Mn and Cr additions [11]. Casting techniques using high cooling rates upon solidification such as high-pressure die casting have also been used to mitigate detrimental IMP morphologies in combination with optimized composition. **Figure 3** schematically shows the effect of an ideal addition of Fe-corrector elements [34].

However, such measures also have their limitations, because they generate increased porosity and sludge formation, decrease machinability [32, 35] and raise the total IMP fraction rather than preventing brittle primary IMPs.

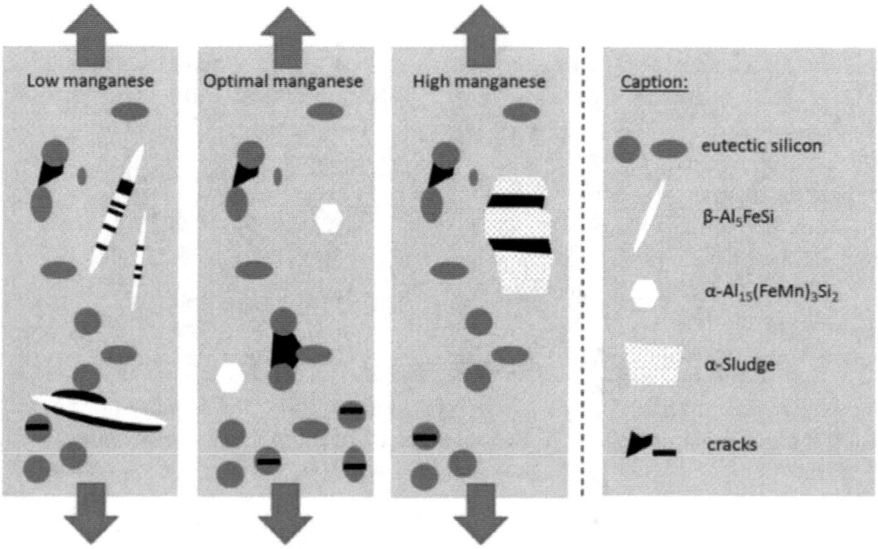

Figure 3: Principal crack initiation in high-pressure die casting alloys with low, optimal and high Mn content as an Fe-corrector. The red arrows mark the loading direction; fewer microcracks (depicted in black) and smaller IMPs are formed when the latter have optimal Mn content. Adapted from [34].

A potential way to influence the IMPs formed is to better understand their nucleation and growth. There is increasing interest in this aspect.

Feng et al. [36] showed via X-ray radiography that TiC and TiB2 addition together with the solidification conditions can refine Fe-IMPs. Approaches to control the properties of Fe-IMPs were also demonstrated in [36–38]. This has been achieved by developing a deeper understanding of of phase selection, nucleation and growth, with the aim of enabling higher Fe content to be tolerated [39–45]. The nuclei of Fe-IMPs were studied in [46]. Minor inoculation effects were reported for α-Al_2O_3, TiC and TiB_2 [40]. Also the role of TiB_2 in the nucleation of Fe-IMPs was studied [45] . Algendy et al. [47] pointed out that even if Mg is not directly incorporated into the Fe-IMPs its content effects their formation. In this context, Que et al. [48] recently showed that the morphology of the formed eutectic Al_6(Fe,Mn) can be modified from from needle- and plate-like to Chinese script via an increase of Mg in Al-1.4Fe-0.7Mn-xMg. Interestingly, they observe the accumulation of Mg on the surface of Al_6(Fe,Mn) particles, with the probable formation of an $Al_{12}Mg_{17}$ layer with a thickness of a few nanometres. The authors conclude that an increasing Mg segregation changes the growth from 2D to 3D [48]. Moreover, grain refinement can positively influence IMPs [46]. In general, the IMPs formed depend strongly on composition, in association with highly complex formation upon solidification involving a series of peritectic reactions [31, 49].

Moreover, in cast alloys a positive effect of increased cooling rates upon solidification is utilized in (for example) high-pressure die casting [34]. It is important to note, that phase selection depends on the solidification conditions. The formed IMPs can be also of a metastable nature with their formation influenced by the growth rate (i.e. the position in an ingot [50]). Sha et al. [41] demonstrated that high growth rates promote the formation of the beneficial IMPs in 6xxx alloys. In a similar context, Wang et al. [51] reported a refinement of Fe-containing IMPs at high speeds during direct chill casting.

In recent years, modified casting techniques have been studied to archive control over IMPs. High shear melt conditioning upon direct chill casting was shown to refine IMPs [52]. Ultrasonication upon casting can also modify the Fe-containing IMPs in a beneficial manner [53]. Further increasing the cooling rate during solidification is worth examining in wrought alloys made by twin roll casting, to promote the formation of fewer harmful Fe-containing IMPs [54]. High-speed twin roll casting has been shown to be beneficial in an Fe-containing 5xxx series alloy [55]. Liu et al. [56] also showed that the cooling rate plays a critical role and that $Al_6(Fe,Mn)$ IMPs can be generated as fine Chinese script morphologies by near-rapid cooling (e.g. continuous strip casting). Moreover, the Fe and Si level plays a critical role [57, 58]. Al-Helal et al. [59] used twin roll casting in combination with high shear melt conditioning. Further methods, such as spray forming can tolerate higher Fe fractions [60].

In the sink cast alloy used for engine blocks in the automotive industry the occurrence of a needle or plate shaped β-AlFeSi phase is common due to slow solidified (1–10 K/s) gravity mold cast components (**Figure 4**). This causes very low elongation to fracture due to the stress concentration effect at the several 100 μm-sized Fe-rich IMPs. On the other side, during additive manufacturing, an extremely high cooling rate of several thousands of K/s is seen. If produced at a high solidification rate, Fe-rich IMPs are <1 μm and show improved elongation values [61, 62]. The difference in IMP size in dependence of the cooling rate is shown in Figure 4.

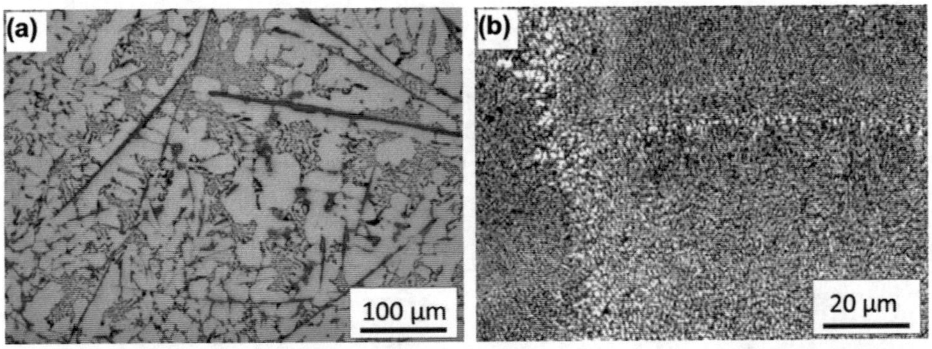

Figure 4. (a) Casting alloy AlSi9Cu3MgFeZn with 1.2% Fe; (a) gravity die casting, (b) additive manufacturing. The size of Fe-rich IMPs is reduced from(a) to (b) from 100 μm below 1μm. Adapted from [61]

Modification of IMP types or morphologies is possible not only during solidification. Heat treatments and deformation have also been used for this purpose in the past [63]. In 6xxx series alloys unfavourable plates or needles have been shown to be transformed by homogenization into less harmful IMPs which could then also be subject to spheroidization [24]. According to Wang et al. [64] this can improve elongation and fatigue strength. Moreover, the surface appearance may be improved because micro surface defects generated by the debonding of IMPs from the Al matrix can be prevented by a suitable heat treatment creating beneficial morphologies [64].
The negative effect of Fe-containing IMPs is also typically reduced by breaking large ones into smaller ones via deformation processes [24]. What is interesting is that negative effects are not always linearly connected with increased Fe content. Lu et al. [65] showed that hemming and bending properties can exhibit complex interdependencies with processing and the chemical composition. The detrimental effect of IMPs can also be mitigated by severe plastic deformation, resulting in a smaller aspect ratio and more randomized distribution, which may improve properties (this is shown for friction stir processing in [66]).

2.3 Heterostructured wrought aluminium alloys containing high fractions of IMPs.

The field of heterostructured materials has attracted increasing research interest in the last few years [67–70]. Although the basic physical principle behind these materials has long been known [71], the recent focus on tailoring strain delocalization to achieve strength-ductility synergy [72] in combination with new metal alloys [73, 74] and novel processing routes ([75, 76]) has given impetus to new views on the topic. Heterogeneous materials are defined as materials with regions of strong difference in strength. In this they have inspired favourable strength-ductility combinations that are not accessible in their homogeneous counterparts [69]. The basis is high back-stress for strengthening and high back-stress strain hardening for good ductility founded on the formation of geometrically necessary dislocations (GNDs). This has also been called hetero-deformation induced strengthening and strain hardening [67]. Note that hetero-deformation induced stress is also known as kinematic stress [67, 77]. Because the back-stress is produced by dislocation pile-up at the interfaces, the number of interfaces needs to be high but with large enough spacing to allow effective dislocation pile-up in such materials [69]. Strain partitioning between parts of the heterostructure should also be considerable, to gain attractive properties [69]. A common industrially applied type is the dual phase structure, i.e. in dual phase steels. There the hard martensitic phase is typically embedded in a soft matrix, and the high strain partitioning results in high strain hardening and good ductility [78]. In additions Kim et al. [79] demonstrated that in steels brittle IMPs can be utilized in analogy to dual phase steel to gain a strong steel with good ductility.

For Aluminium we used fast solidification at a rate of ~60 K/s and thermomechanical processing to control morphology and dispersion to generated an unusually attractive combination of strength and ductility in addition to the substantially increased strain hardening typical of heterostructured materials [80] (see **Figure 5**). At first glance the high ductility found may be unexpected, but is due to attractive distribution and morphology of IMPs, leading to high strain hardening.

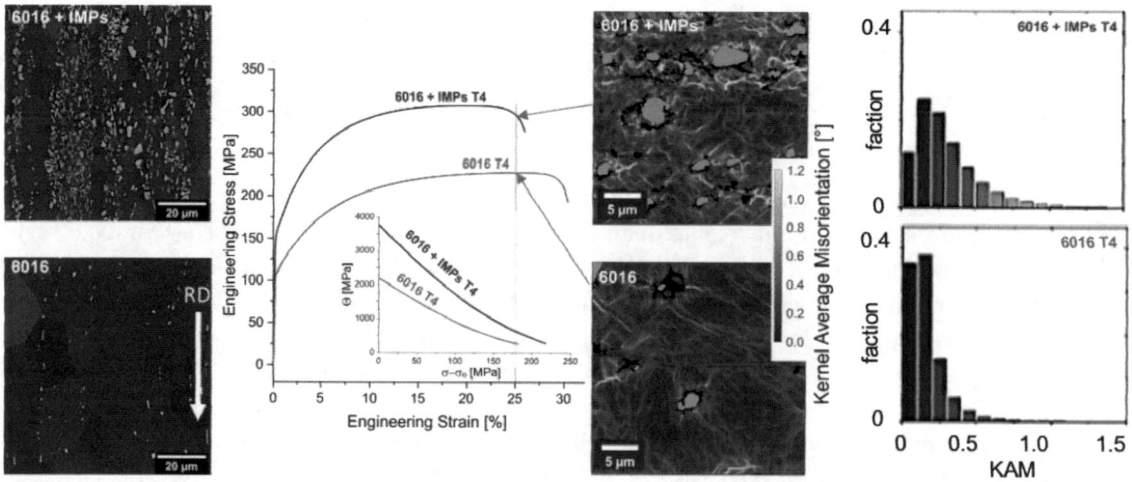

Figure 5: Microstructure and stress-strain curves of commercial alloys 6016 and 6016 doped with 10 vol.% IMPs. The insert shows increased strain hardening due to IMPs. Kernel Average Misorientation (KAM) analysis (right) reveals GNDs around IMPs and a broader distribution of KAM in line with the increased strain hardening for 6016 + IMPs at a strain of 25%. Adapted from [80].

IOP Conf. Series: Materials Science and Engineering **1315** (2024) 012008 doi:10.1088/1757-899X/1315/1/012008

The application of fast solidification is also found to have a beneficial effect on cast alloys. Krall and Pogatscher [81] recently found that a fast solidification of about 60 K/s followed by a two stage homogenization treatment leads to a refinement and spheroidization of the eutectic structures and primary IMPs in secondary cast alloys of the 4xx series. This refinement allows processing of the secondary cast alloys to sheet material by hot and cold rolling, without the occurrence of edge or head cracks. The same procedure is applicable for a range of mixed automotive scrap alloys. Krall et. al. [82] produced alloys with compositions in accordance with alloys precent in the most used passenger cars in combination with different dismantling scenarios (see **Figure 6**). The refined microstructure allows a processing of the mixed automotive scrap as wrought alloy. After solution annealing the so produced alloys exhibit elongation to break in the same region as commercial 5xxx series alloys. Moreover, the alloy allows the option for age hardening through its high Si content in combination with Mg leading to strength values between 200 and 350 MPa, while still retraining an unexceptional high elongation between 19–27% as shown in **Figure 6**. However, it should be mentioned that this is an initial approach and that some further development will be necessary before it can be used for industrial applications. In addition to the currently required high cooling rate during casting, a relatively long homogenization process is also necessary, which requires an increased energy input. Both are the subject of current research efforts.

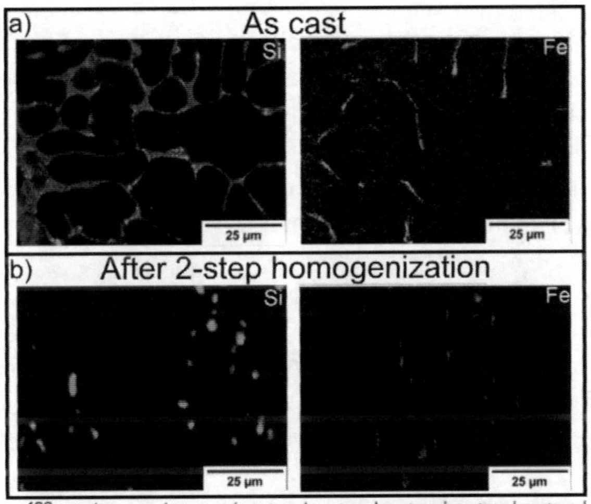

Aluminum alloy compositions calculated for each vehicle type. EU = EU car, PU = pickup, EC = electric car, ND = No dismantling, CD = Current dismantling, TD = Theoretical dismantling, Bal = Balance.

Alloy	Si [%]	Fe [%]	Cu [%]	Mn [%]	Mg [%]	Zn [%]	Al [%]
EU ND	5.16	0.52	0.72	0.34	0.75	0.41	Bal.
EU CD	4.49	0.55	0.67	0.39	0.89	0.44	Bal.
EU TD	3.79	0.47	0.30	0.37	0.89	0.29	Bal.
PU ND	4.45	0.50	1.22	0.25	0.48	0.46	Bal.
PU TD	0.82	0.17	0.12	0.12	0.53	0.11	Bal.
EC ND	1.38	0.24	0.09	0.22	1.45	0.13	Bal.
EC TD	0.60	0.24	0.10	0.23	1.64	0.14	Bal.

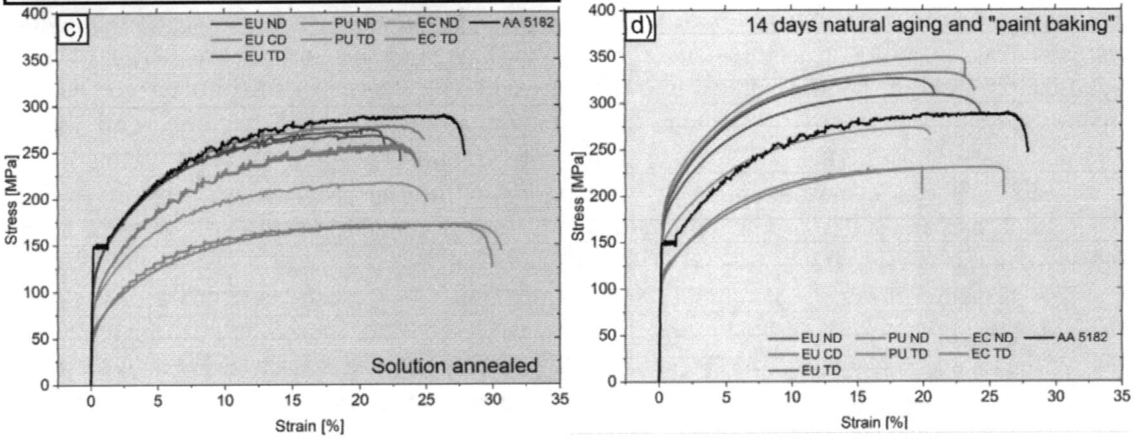

Figure 6: Automotive scrap compositions in different scenarios with varying dismantling operations. IMPs in the as-cast structure (a) can be refined during a two-step homogenization treatment (b). Solution

annealing of 1.2 mm sheet material shows interesting elongation to break values (c) and the alloys have the option for age-hardening processes (d). Adapted from [82]

3. Conclusions and Outlook

With this overview we want to challenge the traditional metallurgical belief that the purity of metals and alloys directly correlates with their mechanical performance. As global production shifts towards sustainable practices, the reliance on recycling materials, particularly aluminium alloys, has highlighted the complexities and opportunities presented by impurities. The transition from viewing impurities as detrimental to harnessing them as integral components of material design marks a significant evolution in material science.

Historically, impurities in aluminium alloys have been perceived as undesirable due to their tendency to form brittle intermetallic phases that compromise material strength and ductility. These phases often manifest as brittle needles or plates, which can initiate cracks and reduce overall ductility and fracture toughness. The traditional approach in metallurgy has focused on minimizing these impurities through refining processes that limit the presence of primary intermetallic phase-forming elements. However, this approach is becoming increasingly unsustainable as the global demand for aluminium rises and recycling becomes more prevalent.

There are already approaches that reimagining impurities as potential assets rather than liabilities. By strategically designing heterostructured alloys, we have developed a concept for a class of upcycled aluminium alloys that can tolerate and even benefit from high levels of impurities. Central to this innovation is the use of intermetallic phases as impurity sinks. These phases, previously viewed as harmful, are transformed into critical components that enhance the mechanical properties of the material by creating a unique heterostructure. This structure achieves a superior balance between strength and ductility, utilizing hetero-deformation-induced strengthening and strain hardening to reach performance levels previously unattainable in conventional alloys.

The key to this transformation lies in the manipulation of intermetallic phases during solidification and solid-state processing. The resulting heterostructures exhibit an optimal balance between strength and ductility, offering a superior combination of mechanical properties compared to traditional homogeneous alloys. This approach will hopefully allow for the design of alloys with good mechanical performance, making them suitable for a wide range of industrial applications.

Future research has to focus on refining processing techniques to ensure the scalability and industrial applicability of heterostructured alloys. This includes optimizing parameters for solidification and thermomechanical treatment to achieve even finer control over intermetallic phase distribution and morphology. Advanced computational modelling and in-situ characterization techniques will further provide deeper insights into the dynamics of intermetallic phase formation and their interactions within the alloy matrix in the next years. These tools can be used to predict and tailor the properties of alloys, paving the way for further advancements in material design.

The transition towards sustainable manufacturing practices is the significant driver in the development of heterostructured alloys. In the context of the automotive industry, where aluminium is extensively used, the ability to efficiently recycle and upcycle scrap is essential. As electric vehicles become more prevalent and the need for cast alloys in engine blocks diminishes, the demand for lightweight, high-performance materials will continue to grow. By reducing dependence on primary aluminium production, these innovations contribute to lower greenhouse gas emissions and energy consumption, aligning with global efforts towards sustainability and environmental conservation. However, the development of these materials also has broad implications for various industrial applications beyond automotive and aerospace. Such heterostructured aluminium alloys may offer new opportunities for innovation and design. The ability to tailor material properties through strategic design of intermetallic phases and heterostructures opens new possibilities for creating customized materials that meet specific performance criteria.

This overview underscores the importance of rethinking traditional approaches to metallurgy in light of modern challenges and opportunities. By leveraging impurities as a resource rather than a hindrance, we can unlock new potentials for materials that are both environmentally and economically viable. The continued advancement of these concepts will be essential in addressing the global demand for sustainable materials and fostering a future where technology and sustainability coexist harmoniously.

4. References

[1] International Aluminium Institute (IAI); 2022. Available from: URL: https://international-aluminium.org.

[2] STATISTA. Global consumption of flat rolled aluminum products in 2020 [cited 2022] Available from: URL: https://www.statista.com/statistics/1115076/share-of-flat-rolled-aluminum-consumption-by-region/.

[3] Capuzzi S, Timelli G. Preparation and Melting of Scrap in Aluminum Recycling: A Review. Metals 2018; 8(4): 249
[https://doi.org/10.3390/met8040249]

[4] Nakajima K, Takeda O, Miki T, Matsubae K, Nakamura S, Nagasaka T. Thermodynamic analysis of contamination by alloying elements in aluminum recycling. Environ Sci Technol 2010; 44(14): 5594–600
[https://doi.org/10.1021/es9038769][PMID: 20536230]

[5] Ostermann F. Anwendungstechnologie Aluminium. Berlin, Heidelberg: Springer Berlin Heidelberg 2014.

[6] European Aluminium [cited 2024] Available from: URL: https://european-aluminium.eu/wp-content/uploads/2022/08/2020-05-13_european-aluminium_circular-aluminium-action-plan_executive-summary.pdf.

[7] Arowosola A, Gaustad G, Brooks L. Aluminum Alloys in Autobodies: Sources and Sinks. In: Chesonis C, editor. Light Metals 2019. Cham: Springer International Publishing 2019; 1381–3.

[8] Modaresi R, Müller DB. The role of automobiles for the future of aluminum recycling. Environ Sci Technol 2012; 46(16): 8587–94
[https://doi.org/10.1021/es300648w][PMID: 22816552]

[9] Arowosola A, Gaustad G. Estimating increasing diversity and dissipative loss of critical metals in the aluminum automotive sector. Resources, Conservation and Recycling 2019; 150: 104382
[https://doi.org/10.1016/j.resconrec.2019.06.016]

[10] Zhu Y, Chappuis LB, Kleine R de, *et al.* The coming wave of aluminum sheet scrap from vehicle recycling in the United States. Resources, Conservation and Recycling 2021; 164: 105208
[https://doi.org/10.1016/j.resconrec.2020.105208]

[11] Apelian D. Aluminum cast alloys: enabling tools for improved performance. Worldwide Report 2009.

[12] Modaresi R, Løvik AN, Müller DB. Component- and Alloy-Specific Modeling for Evaluating Aluminum Recycling Strategies for Vehicles. JOM 2014; 66(11): 2262–71
[https://doi.org/10.1007/s11837-014-0900-8]

[13] Paraskevas D, Ingarao G, Deng Y, Duflou JR, Pontikes Y, Blanpain B. Evaluating the material resource efficiency of secondary aluminium production: A Monte Carlo-based decision-support tool. Journal of Cleaner Production 2019; 215: 488–96
[https://doi.org/10.1016/j.jclepro.2019.01.097]

[14] Martinsen K, Gulbrandsen-Dahl S. Use of Post-consumer Scrap in Aluminium Wrought Alloy Structural Components for the Transportation Sector. Procedia CIRP 2015; 29: 686–91
[https://doi.org/10.1016/j.procir.2015.02.072]

[15] Soo VK, Compston P, Doolan M. Is the Australian Automotive Recycling Industry Heading towards a Global Circular Economy? – A Case Study on Vehicle Doors. Procedia CIRP 2016; 48: 10–5
[https://doi.org/10.1016/j.procir.2016.03.099]

[16] Soo VK, Compston P, Doolan M. Interaction between New Car Design and Recycling Impact on Life Cycle Assessment. Procedia CIRP 2015; 29: 426–31
[https://doi.org/10.1016/j.procir.2015.02.055]
[17] Soo VK, Peeters J, Paraskevas D, Compston P, Doolan M, Duflou JR. Sustainable aluminium recycling of end-of-life products: A joining techniques perspective. Journal of Cleaner Production 2018; 178: 119–32
[https://doi.org/10.1016/j.jclepro.2017.12.235]
[18] van Schaik A, Reuter MA. The use of fuzzy rule models to link automotive design to recycling rate calculation. Minerals Engineering 2007; 20(9): 875–90
[https://doi.org/10.1016/j.mineng.2007.03.016]
[19] Kunal A, Sarita B. Aluminum Market - By Product (Flat Products, Extruded Products, Forged Products, Long Products, Cast Products), Application (Transport, Construction, Packaging, Consumer Durables), Processing Method, Growth Prospects, Regional Outlook & Forecast, 2024 - 2032. Delaware 19975, USA; 2021.
[20] Raabe D, Tasan CC, Olivetti EA. Strategies for improving the sustainability of structural metals. Nature 2019; 575(7781): 64–74
[https://doi.org/10.1038/s41586-019-1702-5][PMID: 31695209]
[21] Gutowski TG, Sahni S, Allwood JM, Ashby MF, Worrell E. The energy required to produce materials: constraints on energy-intensity improvements, parameters of demand. Philos Trans A Math Phys Eng Sci 2013; 371(1986): 20120003
[https://doi.org/10.1098/rsta.2012.0003][PMID: 23359744]
[22] Raabe D, Ponge D, Uggowitzer PJ, *et al.* Making sustainable aluminum by recycling scrap: The science of "dirty" alloys. Progress in Materials Science 2022; 128: 100947
[https://doi.org/10.1016/j.pmatsci.2022.100947]
[23] Ashkenazi D. How aluminum changed the world: A metallurgical revolution through technological and cultural perspectives. Technological Forecasting and Social Change 2019; 143: 101–13
[https://doi.org/10.1016/j.techfore.2019.03.011]
[24] Poznak A, Freiberg D, Sanders P. Automotive Wrought Aluminium Alloys. In: Fundamentals of Aluminium Metallurgy. Elsevier 2018; 333–86.
[25] Canté MV, Lima TS, Brito C, Garcia A, Cheung N, Spinelli JE. An Alternative to the Recycling of Fe-Contaminated Al. J. Sustain. Metall. 2018; 4(3): 412–26
[https://doi.org/10.1007/s40831-018-0188-y]
[26] Sarkar J, Kutty T, Wilkinson D, Embury J, Lloyd D. Tensile properties and bendability of T4 treated AA6111 aluminum alloys. Materials Science and Engineering: A 2004; 369(1-2): 258–66
[https://doi.org/10.1016/j.msea.2003.11.022]
[27] Polmear I, StJohn D, Nie J-F, Qian M. Wrought Aluminium Alloys. In: Light Alloys. Elsevier 2017; 157–263.
[28] Zajac S, Hutchinson B, Johansson A, Gullman L-O. Microstructure control and extrudability of Al–Mg–Si alloys microalloyed with manganese. Materials Science and Technology 1994; 10(4): 323–33
[https://doi.org/10.1179/mst.1994.10.4.323]
[29] Petit T, Besson J, Ritter C, Colas K, Helfen L, Morgeneyer TF. Effect of hardening on toughness captured by stress-based damage nucleation in 6061 aluminum alloy. Acta Materialia 2019; 180: 349–65
[https://doi.org/10.1016/j.actamat.2019.08.055]
[30] Shen Y, Morgeneyer TF, Garnier J, Allais L, Helfen L, Crépin J. Three-dimensional quantitative in situ study of crack initiation and propagation in AA6061 aluminum alloy sheets via synchrotron laminography and finite-element simulations. Acta Materialia 2013; 61(7): 2571–82
[https://doi.org/10.1016/j.actamat.2013.01.035]
[31] Sweet L, Zhu SM, Gao SX, Taylor JA, Easton MA. The Effect of Iron Content on the Iron-Containing Intermetallic Phases in a Cast 6060 Aluminum Alloy. Metall Mater Trans A 2011; 42(7): 1737–49
[https://doi.org/10.1007/s11661-010-0595-6]
[32] Cao X, Campbell J. Morphology of β-Al5FeSi Phase in Al-Si Cast Alloys. MATERIALS TRANSACTIONS 2006; 47(5): 1303–12
[https://doi.org/10.2320/matertrans.47.1303]
[33] Steinbach S, Ratke L, Zimmermann G, Budenkova O. Formation of intermetallic phases in AlSi7Fe1 alloy processed under microgravity and forced fluid flow conditions and their influence on the permeability. IOP Conf. Ser.: Mater. Sci. Eng. 2016; 117: 12019
[https://doi.org/10.1088/1757-899X/117/1/012019]
[34] Bösch D, Pogatscher S, Hummel M, *et al.* Secondary Al-Si-Mg High-pressure Die Casting Alloys with Enhanced Ductility. Metall Mater Trans A 2015; 46(3): 1035–45
[https://doi.org/10.1007/s11661-014-2700-8]

[35] Taylor JA. Iron-Containing Intermetallic Phases in Al-Si Based Casting Alloys. Procedia Materials Science 2012; 1: 19–33
[https://doi.org/10.1016/j.mspro.2012.06.004]
[36] Feng S, Liotti E, Lui A, *et al.* In-situ X-ray radiography of primary Fe-rich intermetallic compound formation. Acta Materialia 2020; 196: 759–69
[https://doi.org/10.1016/j.actamat.2020.06.045]
[37] Feng S, Liotti E, Lui A, *et al.* An in-situ method to estimate the tip temperature and phase selection of secondary Fe-rich intermetallics using synchrotron X-ray radiography. Scripta Materialia 2018; 149: 44–8
[https://doi.org/10.1016/j.scriptamat.2018.02.001]
[38] Wang Y, Jones H. Effect of growth velocity on the growth temperature of the Al-AlxFe eutectic in Al-2.85Fe-0.12V alloy. Metall Mater Trans A 2001; 32(5): 1251–3
[https://doi.org/10.1007/s11661-001-0135-5]
[39] Allen CM, O'Reilly K, Cantor B, Evans PV. Intermetallic phase selection in 1XXX Al alloys. Progress in Materials Science 1998; 43(2): 89–170
[https://doi.org/10.1016/S0079-6425(98)00003-6]
[40] Khalifa W, Samuel FH, GM-NSERC-UQAC, Gruzleski JE, Doty HW, Valtierra S. Nucleation of Fe-intermetallic phases in the Al-Si-Fe alloys. Metall Mater Trans A 2005; 36(4): 1017–32
[https://doi.org/10.1007/s11661-005-0295-9]
[41] Sha G, O'Reilly K, Cantor B, Worth J, Hamerton R. Growth related metastable phase selection in a 6xxx series wrought Al alloy. Materials Science and Engineering: A 2001; 304-306: 612–6
[https://doi.org/10.1016/S0921-5093(00)01545-8]
[42] Sha G, O'Reilly K, Cantor B. Characterization of Fe-Rich Intermetallic Phases in a 6xxx Series Al Alloy. MSF 2006; 519-521: 1721–6
[https://doi.org/10.4028/www.scientific.net/msf.519-521.1721]
[43] Kumar S, Grant PS, O'Reilly KAQ. Evolution of Fe Bearing Intermetallics During DC Casting and Homogenization of an Al-Mg-Si Al Alloy. Metall Mater Trans A 2016; 47(6): 3000–14
[https://doi.org/10.1007/s11661-016-3451-5]
[44] Kumar S, O'Reilly K. Influence of Al grain structure on Fe bearing intermetallics during DC casting of an Al-Mg-Si alloy. Materials Characterization 2016; 120: 311–22
[https://doi.org/10.1016/j.matchar.2016.09.017]
[45] Lui A, Grant PS, Stone IC, O'Reilly KAQ. The Role of Grain Refiner in the Nucleation of AlFeSi Intermetallic Phases During Solidification of a 6xxx Aluminum Alloy. Metall Mater Trans A 2019; 50(11): 5242–52
[https://doi.org/10.1007/s11661-019-05447-y]
[46] Kumar S, Babu NH, Scamans GM, Eskin DG, Fan Z. Solidification behaviour of an AA5754 Al alloy ingot cast with high impurity content. International Journal of Materials Research 2012; 103(10): 1228–34
[https://doi.org/10.3139/146.110760]
[47] Algendy AY, Liu K, Chen X-G. Formation of intermetallic phases during solidification in Al-Mg-Mn 5xxx alloys with various Mg levels. MATEC Web Conf. 2020; 326: 2002
[https://doi.org/10.1051/matecconf/202032602002]
[48] Que Z, Zhou Y, Wang Y, Mendis CL, Fan Z. Effects of Mg addition on the Al6(Fe,Mn) intermetallic compounds and the grain refinement of α-Al in Al-Fe-Mn alloys. Materials Characterization 2021; 171: 110758
[https://doi.org/10.1016/j.matchar.2020.110758]
[49] Sha G, O'Reilly K, Cantor B, Titchmarsh J, Hamerton R. Quasi-peritectic solidification reactions in 6xxx series wrought Al alloys. Acta Materialia 2003; 51(7): 1883–97
[https://doi.org/10.1016/S1359-6454(02)00595-5]
[50] Li YJ, Arnberg L. Solidification structures and phase selection of iron-bearing eutectic particles in a DC-cast AA5182 alloy. Acta Materialia 2004; 52(9): 2673–81
[https://doi.org/10.1016/j.actamat.2004.02.015]
[51] Wang D, Zhang H, Nagaumi H, Li X, Cui J. Microstructural Refinement and α-Dispersoid Evolution in Direct-Chill Cast Al–Mg–Si–Fe Alloy. Adv Eng Mater 2020; 22(12)
[https://doi.org/10.1002/adem.202000517]
[52] K.M. SM, Barekar NS, Lazaro-Nebreda J, Patel JB, Fan Z. In-situ microstructural control of A6082 alloy to modify second phase particles by melt conditioned direct chill (MC-DC) casting process – A novel approach. Journal of Materials Processing Technology 2021; 295: 117170
[https://doi.org/10.1016/j.jmatprotec.2021.117170]
[53] Kotadia HR, Qian M, Das A. Microstructural modification of recycled aluminium alloys by high-intensity ultrasonication: Observations from custom Al–2Si–2Mg–1.2Fe–(0.5,1.0)Mn alloys. Journal of Alloys and Compounds

2020; 823: 153833
[https://doi.org/10.1016/j.jallcom.2020.153833]
[54] Al-Helal K, Chang I, Patel JB, Fan Z. Thermomechanical Treatment of High-Shear Melt-Conditioned Twin-Roll Cast Strip of Recycled AA5754 Alloy. JOM 2019; 71(6): 2018–24
[https://doi.org/10.1007/s11837-018-3190-8]
[55] KOMEDA K, HAGA T. Casting of Recycled A5182 Alloy Strip by High Speed Twin Roll Caster(<Special Issue>M & P 2009 The 9th Materials and Processing Conference). JSMET 2010; 76(766): 694–5
[https://doi.org/10.1299/kikaia.76.694]
[56] Liu Y, Huang G, Sun Y, *et al.* Effect of Mn and Fe on the Formation of Fe- and Mn-Rich Intermetallics in Al-5Mg-Mn Alloys Solidified Under Near-Rapid Cooling. Materials (Basel) 2016; 9(2)
[https://doi.org/10.3390/ma9020088][PMID: 28787888]
[57] Liu Y, Luo L, Han C, Ou L, Wang J, Liu C. Effect of Fe, Si and Cooling Rate on the Formation of Fe- and Mn-rich Intermetallics in Al–5Mg–0.8Mn Alloy. Journal of Materials Science & Technology 2016; 32(4): 305–12
[https://doi.org/10.1016/j.jmst.2015.10.010]
[58] Liu Y, Sun Y, Zhang L, Zhao Y, Wang J, Liu C. Microstructure and Mechanical Properties of Al-5Mg-0.8Mn Alloys with Various Contents of Fe and Si Cast under Near-Rapid Cooling. Metals 2017; 7(10): 428
[https://doi.org/10.3390/met7100428]
[59] Al-Helal K, Patel JB, Fan Z. Fe-Rich Intermetallic Formation and Mechanical Properties of Recycled AA6111 Alloy Strips Produced by Melt Conditioning Twin Roll Casting. JOM 2020; 72(11): 3753–9
[https://doi.org/10.1007/s11837-020-04324-z]
[60] Pereira LH, Asato GH, Otani LB, *et al.* Changing the solidification sequence and the morphology of iron-containing intermetallic phases in AA6061 aluminum alloy processed by spray forming. Materials Characterization 2018; 145: 507–15
[https://doi.org/10.1016/j.matchar.2018.09.006]
[61] Fousova M, Dvorsky D, Vronka M, Vojtech D, Lejcek P. The Use of Selective Laser Melting to Increase the Performance of $AlSi_9Cu_3Fe$ Alloy. Materials (Basel) 2018; 11(10)
[https://doi.org/10.3390/ma11101918][PMID: 30304839]
[62] Fragner W, Pogatscher S, Spierings, AB, Uggowitzer PJ. Advanced alloy development for additive manufacturing. Aluminium 2021; (7): 62–4.
[63] Li Y, Arnberg L. A eutectoid phase transformation for the primary intermetallic particle from Alm(Fe,Mn) to Al3(Fe,Mn) in AA5182 alloy. Acta Materialia 2004; 52(10): 2945–52
[https://doi.org/10.1016/j.actamat.2004.02.041]
[64] Wang Y, Deng Y, Zhang J, Zhao J, Zhang Y. Eliminating microstructural defects and improving the mechanical and fatigue properties of an Al–Mg–Si alloy extrusion by optimising homogenisation. Materials Science and Engineering: A 2019; 764: 138188
[https://doi.org/10.1016/j.msea.2019.138188]
[65] Lu H, Zhang J, Tian N, Song X, Ma M, Lu G. Recycle-Friendly Aluminum Alloy Sheets for Automotive Applications Based on Hemming. Automot. Innov. 2018; 1(1): 70–5
[https://doi.org/10.1007/s42154-018-0012-1]
[66] Hannard F, Castin S, Maire E, Mokso R, Pardoen T, Simar A. Ductilization of aluminium alloy 6056 by friction stir processing. Acta Materialia 2017; 130: 121–36
[https://doi.org/10.1016/j.actamat.2017.01.047]
[67] Zhu Y, Ameyama K, Anderson PM, *et al.* Heterostructured materials: superior properties from hetero-zone interaction. Materials Research Letters 2021; 9(1): 1–31
[https://doi.org/10.1080/21663831.2020.1796836]
[68] Lu L, Wu X, Beyerlein IJ. Preface to the viewpoint set on: Heterogeneous gradient and laminated materials. Scripta Materialia 2020; 187: 307–8
[https://doi.org/10.1016/j.scriptamat.2020.06.036]
[69] Wu X, Zhu Y. Heterogeneous materials: a new class of materials with unprecedented mechanical properties. Materials Research Letters 2017; 5(8): 527–32
[https://doi.org/10.1080/21663831.2017.1343208]
[70] Estrin Y, Beygelzimer Y, Kulagin R, *et al.* Architecturing materials at mesoscale: some current trends. Materials Research Letters 2021; 9(10): 399–421
[https://doi.org/10.1080/21663831.2021.1961908]
[71] Ashby MF. Work hardening of dispersion-hardened crystals. Philosophical Magazine 1966; 14(132): 1157–78
[https://doi.org/10.1080/14786436608224282]

[72] Wu H, Fan G. An overview of tailoring strain delocalization for strength-ductility synergy. Progress in Materials Science 2020; 113: 100675
[https://doi.org/10.1016/j.pmatsci.2020.100675]
[73] Ma E, Wu X. Tailoring heterogeneities in high-entropy alloys to promote strength-ductility synergy. Nat Commun 2019; 10(1): 5623
[https://doi.org/10.1038/s41467-019-13311-1][PMID: 31819051]
[74] Shi P, Li R, Li Y, *et al.* Hierarchical crack buffering triples ductility in eutectic herringbone high-entropy alloys. Science 2021; 373(6557): 912–8
[https://doi.org/10.1126/science.abf6986][PMID: 34413235]
[75] Wang G, Ouyang H, Su Y, *et al.* Heterostructured bulk aluminum with controllable gradient structure: Fabrication strategy and deformation mechanisms. Scripta Materialia 2021; 196: 113762
[https://doi.org/10.1016/j.scriptamat.2021.113762]
[76] Zhang D. Ultrafine grained metals and metal matrix nanocomposites fabricated by powder processing and thermomechanical powder consolidation. Progress in Materials Science 2022; 123: 100853
[https://doi.org/10.1016/j.pmatsci.2021.100853]
[77] Roters F, Eisenlohr P, Hantcherli L, Tjahjanto DD, Bieler TR, Raabe D. Overview of constitutive laws, kinematics, homogenization and multiscale methods in crystal plasticity finite-element modeling: Theory, experiments, applications. Acta Materialia 2010; 58(4): 1152–211
[https://doi.org/10.1016/j.actamat.2009.10.058]
[78] Tasan CC, Diehl M, Yan D, *et al.* An Overview of Dual-Phase Steels: Advances in Microstructure-Oriented Processing and Micromechanically Guided Design. Annu. Rev. Mater. Res. 2015; 45(1): 391–431
[https://doi.org/10.1146/annurev-matsci-070214-021103]
[79] Kim S-H, Kim H, Kim NJ. Brittle intermetallic compound makes ultrastrong low-density steel with large ductility. Nature 2015; 518(7537): 77–9
[https://doi.org/10.1038/nature14144][PMID: 25652998]
[80] Trink B, Weißensteiner I, Uggowitzer PJ, Strobel K, Pogatscher S. High Fe content in Al-Mg-Si wrought alloys facilitates excellent mechanical properties. Scripta Materialia 2022; 215: 114701
[https://doi.org/10.1016/j.scriptamat.2022.114701]
[81] Krall P, Pogatscher S. Investigating the Potential of Secondary Aluminum Cast Alloys Used as Wrought Alloys. In: Wagstaff S, editor. Investigating the Potential of Secondary Aluminum Cast Alloys Used as Wrought Alloys; 2024. Cham: Springer; 191–5.
[82] Krall P, Weißensteiner I, Pogatscher S. Recycling aluminum alloys for the automotive industry: Breaking the source-sink paradigm. Resources, Conservation and Recycling 2024; 202: 107370
[https://doi.org/10.1016/j.resconrec.2023.107370]

Acknowledgments

Funded/Co-funded by the European Union (ERC, HETEROCIRCAL, 101124514). Views and opinions expressed are however those of the author(s) only and do not necessarily reflect those of the European Union or the European Research Council. Neither the European Union nor the granting authority can be held responsible for them.

AUTHOR INDEX

Institute of Physics
Dirac House, Temple Back
Bristol BS1 6BE UK

ISSN: 1757-8981
ISBN 979-8-3313-0662-5
9 798331 306625

CURRAN ASSOCIATES INC.
proceedings
.com

International Nuclear Science Technology and Engineering Conference (iNuSTEC2024)

IOP Conference Series: Materials Science and Engineering
Volume 1326

Selangor, Malaysia
14-16 November 2024

Institute of Physics